T0401114

ENERGY SCIENCE, ENGINEERING AND TECHNOLOGY

ETHANOL BIOFUEL PRODUCTION

ENERGY SCIENCE, ENGINEERING AND TECHNOLOGY

Oil Shale Developments
Ike S. Bussell (Editor)
2009. ISBN: 978-1-60741-475-9

Power Systems Applications of Graph Theory
Jizhong Zhu
2009. ISBN: 978-1-60741-364-6

Bioethanol: Production, Benefits and Economics
Jason B. Erbaum (Editor)
2009. ISBN: 978-1-60741-697-5

Bioethanol: Production, Benefits and Economics
Jason B. Erbaum (Editor)
2009. ISBN: 978-1-61668-000-8 (Online Book)

Introduction to Power Generation Technologies
Andreas Poullikkas
2009. ISBN: 978-1-60876-472-3

Handbook of Exergy, Hydrogen Energy and Hydropower Research
Gaston Pélissier and Arthur Calvet (Editors)
2009. ISBN: 978-1-60741-715-6

Energy Costs, International Developments and New Directions
Leszek Kowalczyk and Jakub Piotrowski (Editors)
2009. ISBN: 978-1-60741-700-2

Radial-Bias-Combustion and Central-Fuel-Rich Swirl Pulverized Coal Burners for Wall-Fired Boilers
Zhengqi Li
2009. ISBN: 978-1-60876-455-6

Syngas Generation from Hydrocarbons and Oxygenates with Structured Catalysts
Vladislav Sadykov, L. Bobrova, S. Pavlova, V. Simagina, L. Makarshin, V. Julian, R. H. Ross, and Claude Mirodatos
2009 ISBN: 978-1-60876-323-8

Corn Straw and Biomass Blends: Combustion Characteristics and NO Formation
Zhengqi Li
2009. ISBN: 978-1-60876-578-2

Computational Techniques: The Multiphase CFD Approach to Fluidization and Green Energy Technologies (includes CD-ROM)
Dimitri Gidaspow and Veeraya Jiradilok
2009. ISBN: 978-1-60876-024-4

Air Conditioning Systems: Performance, Environment and Energy Factors
Tobias Hästesko and Otto Kiljunen (Editors)
2009. ISBN: 978-1-60741-555-8

Cool Power: Natural Ventilation Systems in Historic Buildings
Carla Balocco and Giuseppe Grazzini
2010. ISBN: 978-1-60876-129-6

A Sociological Look at Biofuels: Understanding the Past/Prospects for the Future
Michael S. Carolan
2010: ISBN: 978-1-60876-708-3

Ethanol Biofuel Production
Bratt P. Haas (Editor)
2010: ISBN: 978-1-60876-086-2

ENERGY SCIENCE, ENGINEERING AND TECHNOLOGY

ETHANOL BIOFUEL PRODUCTION

BRATT P. HAAS
EDITOR

Nova Science Publishers, Inc.
New York

For permission to use material from this book please contact us:
Telephone 631-231-7269; Fax 631-231-8175
Web Site: http://www.novapublishers.com

LIBRARY OF CONGRESS CATALOGING-IN-PUBLICATION DATA

Ethanol biofuel production / editor, Bratt P. Haas.
p. cm.
Includes index.
ISBN 978-1-60876-086-2 (hardcover)
1. Ethanol as fuel. 2. Biomass energy. I. Haas, Bratt P.
TP339.E84 2009
662'.6692--dc22
2009041027

Published by Nova Science Publishers, Inc. ✦ New York

CONTENTS

PREFACE

Fuel ethanol production has increased steadily in the U.S. since the 1980's, when it was given impetus by the need to reduce energy dependence on foreign supplies. The momentum has continued as production costs have fallen, and as the U.S. Clean Air Act has specified a percentage of renewable fuels to be mixed with gasoline. New technologies that may further increase cost savings include coproduct development, such as recovery of high-value food supplements, and cellulosic conversion. Though improvements in processing and technology are important, however, the fluctuating rise of inputs such as corn, the cost of energy alternatives, and environmental developments play larger roles in the fortunes of the industry. This book examines the use of ethanol as fuel, as well as its other applications in different parts of the world. This book also addresses a policy initiative by the Federal Administration to apply United States Department of Energy (DOE) research to broadening the country's domestic production of economic, flexible, and secure sources of energy fuels. This book consists of public documents which have been located, gathered, combined, reformatted, and enhanced with a subject index, selectively edited and bound to provide easy access.

Chapter 1 - Fuel ethanol production has increased steadily in the United States since the 1980s, when it was given impetus by the need to reduce energy dependence on foreign supplies. The momentum has continued as production costs have fallen, and as the U.S. Clean Air Act has specified a percentage of renewable fuels to be mixed with gasoline. The fraction of annual U.S. corn production used to make ethanol rose from around 1 percent in 1980 to around 20 percent in 2006, and ethanol output rose from 175 million gallons to about 5.0 billion gallons over the same period. New technologies that may further increase cost savings include coproduct development, such as recovery of high-value food supplements, and cellulosic conversion. High oil prices may spur the risk-taking needed to develop cellulose-to-ethanol production. Developments such as dry fractionation technology, now commercially viable, may alter the structure of the industry by giving the cheaper dry-grind method an edge over wet milling. Dry milling requires smaller plants, and local farmer cooperatives could flourish as a result. Though improvements in processing and technology are important, however, the fluctuating price of inputs such as corn, the cost of energy alternatives, and environmental developments play larger roles in the fortunes of the industry.

Chapter 2 - One of our greatest challenges is to reduce our nation's dependence on imported petroleum. To accomplish this, we need a variety of alternative fuels, including ethanol produced from cellulosic materials like grasses and wood chips. Fortunately, the United States has abundant agricultural and forest resources that can be converted into biofuels. Recent studies by the U.S. Department of Energy (DOE) suggest these resources can be used to produce enough ethanol – 60 billion gallons/year – to displace about 30% of our current gasoline consumption by 2030.

Chapter 3 - This chapter addresses a policy initiative by the Federal Administration to apply United States Department of Energy (DOE) research to broadening the country's domestic production of economic, flexible, and secure sources of energy fuels. President Bush stated in his 2006 State of the Union Address: "America is addicted to oil." To reduce the Nation's future demand for oil, the President has proposed the Advanced Energy Initiative which outlines significant new investments and policies to change the way we fuel our vehicles and change the way we power our homes and businesses. The specific goal for biomass in the Advanced Energy Initiative is to foster the breakthrough technologies needed to make cellulosic ethanol cost-competitive with corn-based ethanol by 2012.

Chapter 4 - Pertinent information regarding whey-to-fuel ethanol production is explored and reviewed. A potential of producing up to 203 million gallons of fuel ethanol from whey in 2006 was estimated, and dairy cooperatives could have a share of 65 million gallons. Two whey-ethanol plants are currently operated by dairy cooperatives, producing a total of 8 million gallons a year. Successful operations of the plants since the 1980s indicate that (1) fuel ethanol production from whey is technically feasible, (2) whey-to-fuel ethanol production technologies and processes are mature and capable of being adopted for commercial operations, and (3) producing fuel ethanol from whey is economically feasible. However, in this era of whey products' price uncertainties, a key consideration in assessing the feasibility of a new whey-ethanol venture should be the valuation of the opportunity cost of whey as feedstock for fermentation. A new whey- ethanol plant probably should have an annual production capacity of at least 5 million gallons of ethanol. Some historical lessons on the pitfalls to avoid are summarized.

In: Ethanol Biofuel Production
Editor: Bratt P. Haas
ISBN: 978-1-60876-086-2

Chapter 1

NEW TECHNOLOGIES IN ETHANOL PRODUCTION

C. Matthew Rendleman and Hosein Shapouri

ABSTRACT

Fuel ethanol production has increased steadily in the United States since the 1980s, when it was given impetus by the need to reduce energy dependence on foreign supplies. The momentum has continued as production costs have fallen, and as the U.S. Clean Air Act has specified a percentage of renewable fuels to be mixed with gasoline. The fraction of annual U.S. corn production used to make ethanol rose from around 1 percent in 1980 to around 20 percent in 2006, and ethanol output rose from 175 million gallons to about 5.0 billion gallons over the same period. New technologies that may further increase cost savings include coproduct development, such as recovery of high-value food supplements, and cellulosic conversion. High oil prices may spur the risk-taking needed to develop cellulose-to-ethanol production. Developments such as dry fractionation technology, now commercially viable, may alter the structure of the industry by giving the cheaper dry-grind method an edge over wet milling. Dry milling requires smaller plants, and local farmer cooperatives could flourish as a result. Though improvements in processing and technology are important, however, the fluctuating price of inputs such as corn, the cost of energy alternatives, and environmental developments play larger roles in the fortunes of the industry.

ACKNOWLEDGMENTS

The authors wish to thank a number of people who made valuable suggestions and corrections to the paper. They include Don Erbach and Andrew McAloon of the Agricultural Research Service, USDA, Jack Huggins of the Nature Conservancy, and Vijay Singh of the Dept. of Engineering at the University of Illinois at Urbana-Champaign.

About the Authors

C. Matthew Rendleman is with the Dept. of Agribusiness Economics, Southern Illinois University, and Hosein Shapouri is with the Office of Energy Policy and New Uses, USDA.

Introduction

The use of ethanol for fuel was widespread in Europe and the United States until the early 1900s (Illinois Corn Growers' Association/Illinois Corn Marketing Board). Because it became more expensive to produce than petroleum-based fuel, especially after World War II, ethanol's potential was largely ignored until the Arab oil embargo of the 1970s. One response to the embargo was increased use of the fuel extender "gasohol " (or E-10), a mixture of one part ethanol made from corn mixed with nine parts gasoline. Because gasohol was made from a renewable farm product, it was seen in the United States as a way to reduce energy dependence on foreign suppliers.

After the oil embargo ended, the use of ethanol increased, even though the price of oil fell and for years stayed low. Ethanol became cheaper to make as its production technology advanced. Agricultural technology also improved, and the price of corn dropped. By 1992, over 1 billion gallons of fuel ethanol were used annually in the United States, and by 2004 usage had risen to over 3.4 billion gallons. Many farm groups began to see ethanol as a way to maintain the price of corn and even to revitalize the rural economy. This economic support for ethanol coincided with a further justification for its use: to promote clean air. A 10-percent ethanol mixture burns cleaner than gasoline alone (reducing the emission of particulate matter, carbon monoxide, and other toxins), giving ethanol a place in the reformulated gasoline (RFG) market.

Ethanol use has also been boosted by the U.S. Clean Air Act and its various progressions. Originally, the Clean Air Act required wintertime use of oxygenated fuels in some urban areas to ensure more complete burning of petroleum fuels. Since ethanol contains 35 percent oxygen, this requirement of the act could be met by using an ethanol-containing blend. The current Energy Act eliminates the need for oxygenates per se in RFG, but it specifies the minimum amount of renewable fuels to be added to gasoline.

By 1980, fuel ethanol production had increased from a few million gallons in the 1970s to 175 million gallons per year. During the 1990s, production increased to 1.47 billion gallons, and total production for 2006 is expected to be about 5.0 billion gallons. Annual U.S. plant capacity is now over 4.5 billion gallons, most of it currently in use. Demand is rising partly because a number of States have banned (or soon will ban) methyl tertiary-butyl ether (MTBE), and ethanol is taking over MTBE's role (Dien et al., April 2002). Ethanol provides a clean octane replacement for MTBE. The California Energy Commission and the California Department of Food and Agriculture now support ethanol development, and ethanol's use in California alone is expected to reach 1.25 billion gallons by 2012 (Ross).

The fraction of the Nation's annual corn production used to make ethanol rose from around 6.6 million bushels in the early 1980s (1 percent) to approximately 2 billion bushels in 2006 (20 percent) (figure 1).

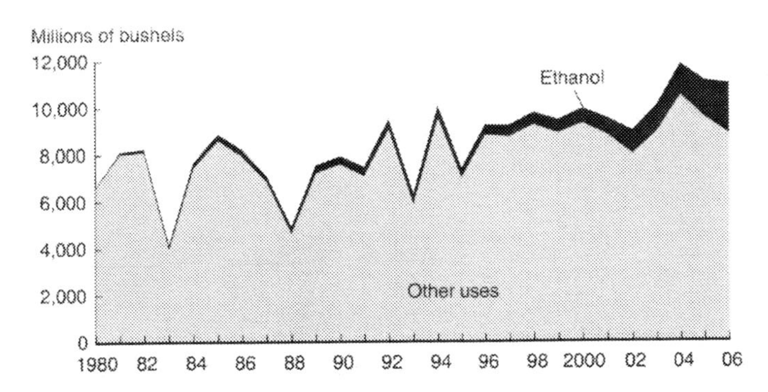

Source: *Feed Situation and Outlook Yearbook*, USDA, Economic Research Service, various years.

Figure 1. Ethanol's use of U.S. corn production

CHANGES SINCE THE 1993 ERS ANALYSIS OF ETHANOL PRODUCTION

In 1993, USDA's Economic Research Service (ERS) published *Emerging Technologies in Ethanol Production,* a report on the then-current state of ethanol production technology and efficiency (Hohmann and Rendleman). The report included a summary of production costs (table 1) and predictions of "near-term" and "long-term" technological advances that many believed would bring down ethanol costs.

The numbers were based on the costs of wet milling, which was then by far the greatest source of output. (Milling types are explained in the next section.) The estimate included a capital cost component, which distinguished this estimate from others done at the time. Other estimates ranged from $1.08 to $1.95 per gallon.

The near-term technologies listed in the ERS report were as follows:

- Gaseous injection of sulfur dioxide and the use of special corn hybrids,
- Membrane filtration,
- Other advances, including improved yeast strains and immobilization of yeast in gel substrates.

Long-term technologies (potentially available in 5 to 10 years) were as follows:

- Bacterial fermentation,
- Conversion of corn fiber to ethanol (cellulosic conversion),
- Coproduct development.

Though the savings from technological improvements are significant, they tend to be small compared with fluctuations in the net cost of corn, the main ethanol feedstock. This is illustrated in table 2, which presents data on corn costs and profits from the coproduct DDGS (distiller's dried grains with solubles) in dry-mill ethanol production from 1981-2004. Since 1981, sales of DDGS have recovered nearly half the cost of each bushel of corn used to produce ethanol, peaking in 1986, when over 66 percent of the feedstock cost was recovered this way. In recent years, the percentage of recovery has fallen because increased demand for ethanol has led to an abundance of DDGS, lowering its price on the feed market.

Table 1. Ethanol Wet-Mill Cost Estimates, 1993

Cost category	Cost per gallon
Feedstock	$0.44
Capital	$0.43
Operating	$0.37
Total	$1.24

Source: Hohmann and Rendleman (1993).

Table 2. Net Corn Costs of Dry Milling,1981-2004[1]

Year	Corn value	DDGS* value	Byproduct value	Net corn cost	Net corn cost
	$/bu	*$/bu*	*% corn cost*	*$/bu*	*$/gal. ethanol*
1981	2.47	1.25	50.5	1.22	0.45
1982	2.55	1.21	47.5	1.34	0.50
1983	3.21	1.47	45.7	1.74	0.65
1984	2.63	0.83	31.6	1.80	0.67
1985	2.23	0.92	41.1	1.31	0.49
1986	1.50	1.00	66.4	0.50	0.19
1987	1.94	1.16	59.9	0.78	0.29
1988	2.54	1.20	47.2	1.34	0.50
1989	2.36	1.06	44.8	1.30	0.48
1990	2.28	1.08	47.3	1.20	0.45
1991	2.37	1.04	43.9	1.33	0.49
1992	2.07	1.04	50.4	1.03	0.38
1993	2.50	1.05	42.1	1.45	0.54
1994	2.26	0.91	40.1	1.35	0.50
1995	3.24	1.29	39.7	1.95	0.72
1996	2.71	1.21	44.8	1.50	0.55
1997	2.43	0.93	38.4	1.50	0.55
1998	1.94	0.72	37.2	1.22	0.45
1999	1.82	0.68	37.3	1.14	0.42
2000	1.85	0.69	37.0	1.16	0.43
2001	1.97	0.68	34.6	1.29	0.48
2002	2.32	0.75	32.3	1.57	0.58
2003	2.42	0.98	40.6	1.44	0.53
2004	2.06	0.65	31.3	1.41	0.52
Average	2.32	0.99	43.0	1.33	0.49

*Distiller's dried grains with solubles.

[1]2.7 gal. of ethanol and 17 lbs of DDGS per bushel of corn.

Source: ERS Feedgrains Database, http://www.ers.usda.gov/db/feedgrains.

The near- and long-term technologies listed in the 1993 ERS analysis were predicted to save from 5 to 7 cents per gallon in the short term and from 9 to 15 cents by 2001. The savings have been as anticipated, but they have not come in the manner predicted.

Gaseous injection of sulfur dioxide was beginning in 1993 and is a part of the quick-germ (QG) and quick-fiber (QF) techniques currently being developed. There is revived interest in the use of special corn hybrids high in starch, though their use is still not widespread. Membrane filtration and yeast immobilization were being used in some plants in 1993, but their use, contrary to expectations, has not increased. Bacterial fermentation is still not used commercially, nor is cellulosic conversion of corn fiber. There have been no outstanding developments in coproducts, but the potential remains for their future exploitation. Most of the cost savings have been through plant automation and optimization of existing processes.

The industry is still improving technologically. It is far more mature than in 1993, and new developments appear poised to bring costs down further and to reduce the environmental impact of producing ethanol. In this report, we examine various production technologies, beginning with input improvements and then discussing process improvements, environmental technologies, and technologies involving coproducts. Finally, we look at niche markets and briefly examine cellulosic conversion.

ETHANOL'S ENERGY EFFICIENCY

Improvements in ethanol's energy consumption have continued since large-scale commercial production began in the 1970s. The process has become more efficient at using the starch in the corn kernel, approaching the theoretical limit of about 2.85 gallons of ethanol per bushel. Energy for conversion has fallen from as high as 70,000 Btu's per gallon in the late 1970s (Wang, August 1999) to 40,000 Btu's or less for modern dry mills and to 40,000-50,000 Btu's for wet mills. Modern energy-saving technology and process optimization account for the improvement.

In 2002, Shapouri et al. surveyed energy values and reported that fuel ethanol from corn produced about 34 percent more energy than it took to produce it. That figure was based on a weighted average of a 37-percent increase in energy from ethanol produced in dry mills and a 30-percent increase from wet mills. This value was revised in 2004 by updating energy estimates for corn production and yield, improving estimates of energy required to produce nitrogen fertilizer and energy estimates for seed corn, and using better methodologies for allocating energy for producing coproducts. With these revisions, the energy gain is 57 percent for wet milling and 77 percent for dry milling, yielding a new weighted average of 67 percent.

The energy content, however, may be less important than the energy replaced. A gallon of ethanol can save 26,575 Btu's of energy by replacing a gallon of gasoline because of ethanol's higher combustion efficiency (Levelton Engineering Ltd. and (S&T)2 Consulting Inc.). A gallon of ethanol containing 76,330 Btu's is able to replace a gallon of gasoline containing about 115,000 Btu's because ethanol's higher octane rating (113-115, compared with 87) allows high-compression engines to perform as well with fewer Btu's.

Ethanol Production Processes

Though new technology may eventually blur the distinction between them, ethanol is produced by one of two processes: wet milling and dry milling. Wet mills are more expensive to build, are more versatile in terms of the products they can produce, yield slightly less ethanol per bushel, and have more valuable coproducts. Wet milling initially accounted for most of the ethanol fuel production in the United States, but new construction has shifted to dry mills, partly because dry mills cost less to build. In 2004, 75 percent of ethanol production came from dry mills and only 25 percent from wet mills (Renewable Fuels Association). As a result, most new technologies are being developed for dry-mill production.

Dry-milling plants have higher yields of ethanol; a new plant can produce 2.8 gallons per bushel, compared with about 2.7 gallons for wet mills. The wet mill is more versatile, though, because the starch stream, being nearly pure, can be converted into other products (for instance, high-fructose corn syrup (HFCS)). Coproduct output from the wet mill is also more valuable.

In each process, the corn is cleaned before it enters the mill. In the dry mill, the milling step consists of grinding the corn and adding water to form the mash. In the wet mill, milling and processing are more elaborate because the grain must be separated into its components. First, the corn is steeped in a solution of water and sulfur dioxide (SO_2) to loosen the germ and hull fiber. This 30- to 40-hour extra soaking step requires additional tanks that contribute to the higher construction costs. Then the germ is removed from the kernel, and corn oil is extracted from the germ. The remaining germ meal is added to the hulls and fiber to form the corn gluten feed (CGF) stream. Gluten, a high-protein portion of the kernel, is also separated and becomes corn gluten meal (CGM), a high-value, high-protein (60 percent) animal feed. The corn oil, CGF, CGM, and other products that result from the production of ethanol are termed coproducts.

Unlike in dry milling, where the entire mash is fermented, in wet milling only the starch is fermented. The starch is then cooked, or liquefied, and an enzyme added to hydrolyze, or segment, the long starch chains. In dry milling, the mash, which still contains all the feed coproducts, is cooked and an enzyme added. In both systems a second enzyme is added to turn the starch into a simple sugar, glucose, in a process called saccharification. Saccharification in a wet mill may take up to 48 hours, though it usually requires less time, depending on the amount of enzyme used. In modern dry mills, saccharification has been combined with the fermentation step in a process called simultaneous saccharification and fermentation (SSF).

Glucose is then fermented into ethanol by yeast (the SSF step in most dry-milling facilities). The mash must be cooled to at least 95° F before the yeast is added. The yeast converts the glucose into ethanol, carbon dioxide (CO_2), and small quantities of other organic compounds during the fermentation process. The yeast, which produces almost as much CO_2 as ethanol, ceases fermenting when the concentration of alcohol is between 12 and 18 percent by volume, with the average being about 15 percent (Shapouri and Gallagher). An energy-consuming process, the distillation step, is required to separate the ethanol from the alcohol-water solution. This two-part step consists of primary distillation and dehydration. Primary distillation yields ethanol that is up to 95-percent free of water. Dehydration brings the concentration of ethanol up to 99 percent. Finally, gasoline is added to the ethanol in a step called "denaturing," making it unfit for human consumption when it leaves the plant.

The coproducts from wet milling are corn oil and the animal feeds corn gluten feed (CGF) and corn gluten meal (CGM). Dry milling production leaves, in addition to ethanol, distiller's dried grains with solubles (DDGS). The feed coproducts must be concentrated in large evaporators and then dried. The CO_2 may or may not be captured and sold.

INPUT IMPROVEMENTS: HIGHER-ETHANOL-YIELDING CORN

Efficient ethanol plants can convert 90-97 percent of the corn's starch content to ethanol. However, not all batches of corn leave the same amount of starch residue. Studies of ethanol yields from different batches show significant variability (Dien et al., March 2002). Even though it is the starch that is turned into ethanol, researchers have been unable to find a correlation between starch content (or even starch extractability) and the final yield of ethanol (Singh and Graeber). Researchers believe some starches are in a more available form (Dien et al., March 2002). They do not know, however, what makes the starch break down easily to simple sugars and why this trait varies from hybrid to hybrid (Bothast, quoted in Bryan, 2002). Some research shows that although the ease with which the starch breaks down varies among hybrids, most of the variability in the breakdown is due to other factors (Haefele et al.).

Seed companies like Pioneer, Monsanto, and Syngenta are working to create corn that will boost ethanol yield. "Current work centers on identifying highly fermentable hybrids we already have," says a Monsanto spokesman (Krohn). Pioneer reports yield increases of up to 6 percent in batches using what it calls HTF corn (High Total Fermentables), compared with the yields from unselected varieties (Haefele et al.). Monsanto calls its selected varieties HFC for High Fermentable Corn.

Syngenta Seeds' Gary Wietgrefe points out several of the impediments to widespread adoption of HTF corn (Ed Zdrojewski in *BioFuels Journal*, 2003b). To begin with, starch and ethanol yield vary by geographic region and from year to year, making an optimizing hybrid choice difficult. Further, choosing a hybrid that maximizes ethanol qualities may mean a tradeoff with yield and other potentially valuable qualities, such as protein content and even test weight (because of moisture). Testing equipment presents its own challenges. Not all units are calibrated the same, creating uncertainty. The technology may be less available to farmers than to ethanol plants, and even if it were to become more readily available, sorting grain by starch availability and making marketing decisions would be problematic for the grower. Finally, with ethanol plants already able to convert between 90 and 97 percent of the corn's starch, any new HTF or HFC genetics or technology would have to overcome the problems and significantly improve the profits of the ethanol plant and the farmer.

Higher ethanol yield leaves less DDGS for animal feed, possibly changing the quality as well as the quantity of the feed coproduct. A lower quantity might raise the protein percentage, but it could also concentrate some of the undesirable contents of the DDGS. Any changes, however, are expected to be minor. (See Haefele et al., p.14, on the selection of hybrids.)

Unlike many technologies that are adopted because they show an immediate improvement in profit or a reduction in risk, corn with a higher ethanol yield does not necessarily lead to additional profits for the farmer in today's marketing environment. Corn is not graded on the basis of fermentability, nor is a premium offered in the wider marketplace for this trait. In order to overcome this market drawback, companies like Monsanto and Pioneer are developing programs to encourage the adoption of their selected hybrids.

So far, hybrid-testing research has centered on dry-mill production, the lower investment technique of choice for the new cooperatively owned plants. The seed companies are targeting their incentive programs on dry mills. Monsanto's program, "Fuel Your Profits," provides the participating ethanol plant with high-tech equipment that profiles the genetics of incoming corn and is calibrated to maximize ethanol yield (Rutherford). As an incentive, Monsanto gives rebates on E85 vehicles (those designed to run on 85%-ethanol fuel) and fueling stations. Pioneer has developed a whole-grain Near Infrared Test (NIT) to identify ethanol yield potential quickly.

PROCESS IMPROVEMENTS

New construction today is mostly of dry mills, and most new technology is designed for them. Major technology changes are made more efficiently while a plant is being built than when they are adopted later.

Advances in Separation Technologies

New techniques that separate corn kernel components before processing will blur the distinction between wet and dry milling (dry grind) by allowing the dry mill to recover the coproducts from the germ. Process improvements are also being made that will reduce the cost of wet milling, generally by shortening the soaking step. Some of the separation improvements we describe here, though promising, are still experimental.

Germ and Fiber Separation

Modifications of the dry-grind facility have made the recovery of corn germ possible in dry milling. Normally, neither corn germ nor any other corn fraction is separated out before becoming part of the mash; all components go through fermentation and become part of the feed coproduct, DDGS. Various modifications of the process have made it possible to recover fiber and corn germ—and thus corn oil—from both the endosperm and the pericarp (outer covering) of the kernel.

A technique developed at the University of Illinois called Quick Germ (QG) allows recovery of corn oil and corn germ meal from the germ, making the dry mill a more profitable operation (Singh and Eckhoff, 1996, 1997; Taylor et al., 2001; Eckhoff, 2001). Results published in 1995 and 1996 demonstrated that with a 3- to 6-hour soak step (as opposed to 24 to 48 hours for the soak step in wet milling), the corn germ could be removed. Since then, research has explored the parameters of the process and its savings potential (Singh and Johnston). Another process, Quick Fiber (QF), can be used with QG to recover fiber from the pericarp, a source of potentially valuable food coproducts. Though these processes have not been used in commercial applications, they hold promise for reducing the net cost of the input corn. The tanks and equipment for the additional steps would increase the plant's capital cost, but could increase its capacity by reducing the amount of nonfermentables in the mash.

Enzymatic Dry Milling

This process, which uses newly developed enzymes, is another method with the potential for cost savings. In addition to recovering the germ and fiber from the pericarp, it allows recovery of endosperm fiber. Savings come from the recovered coproducts and from reduced energy consumption—the process requires less heat for liquefaction and saccharification. Plant capacity should be enhanced as well, since there is less nonfermentable material in the substrate because it is removed earlier in the process. The amount of DDGS is smaller, but of higher quality. Ethanol concentrations in the mash are also higher. Industrial- and food-grade products can be recovered from the fiber. Alternatively, the fiber can be fermented.

Dry Fractionation

This recent technology separates the corn kernel into its components without the soaking step. Depending on the process—several companies currently offer similar technologies—the feedstock may be misted with water before being separated into bran, germ, and the high-starch endosperm portion of the kernel (*BioFuels Journal,* 2005d,e).

The advantages of dry fractionation over processes that require a soak step are threefold: lower costs because less energy is required for drying the feed coproduct, lower emissions, and greater coproduct output because the mash is more highly concentrated. The germ can be sold or pressed for corn oil, and the bran also has potential for food or energy use.

Dry fractionation is a process that has been tested and is in use in the food industry (Madlinger). Both new and planned ethanol plant construction employ the technology. Unlike some other new technologies, the dry fractionation equipment can be added to an existing dry mill.

With all the separation techniques, there appears to be less total ethanol recovered per bushel than with conventional dry-milling techniques, probably due to removal of some starch with the coproducts (Singh and Johnston). Each technique will change the nature of the resulting distiller's grains, potentially raising their value due to a higher protein content; the feed coproduct from the separation processes is purported to be higher in protein and lower in fiber than ordinary DDGS. However, research is needed to determine the feed value of this altered coproduct. Preliminary feed trials with poultry and hogs, as yet unpublished, are promising (Madlinger).

Ammoniation Process in the Wet Mill

Researchers have also investigated a separation technique involving pretreatment with ammonia (Taylor et al., 2003). This process would facilitate removal of the pericarp and reduce the soak time in wet milling or the QG process. Anhydrous ammonia would take the place of the caustic soda solution usually used in debranning. In laboratory research, the pericarp was more easily removed through ammoniation, but though the oil was not degraded, its quantity was reduced compared with conventional techniques.

Continuous Membrane Reactor for Starch Hydrolysis

This process, still experimental, uses enzymatic saccharification of liquefied corn in a membrane reactor. In a continuous membrane reactor, as opposed to the traditional batch process, starch would be broken down and glucose extracted continuously. Theoretically, the yield would increase, and the automated continuous process would enable better control than the batch process.

Alkali Wet Milling

In an experimental modification of the wet-milling process, corn was soaked briefly in sodium hydroxide (NaOH) and debranned (Eckhoff et al.). This process cut the costly soaking time to 1 hour. The pericarp removed in alkali wet milling becomes a potentially valuable part of the coproduct stream. Additional work is needed to develop ways of disposing of or recycling the NaOH before the technique can be commercialized.

New Ways of Fermentation

High-Gravity Fermentation

This technique, still experimental, would lower water use in ethanol production. Potential savings would come from the reduced cost of water and wastewater cleanup, as well as from reduced energy use. This process would involve less heating and cooling per gallon of ethanol. Very-high-gravity fermentation accomplishes this saving in energy by using a highly concentrated mash with more than 30 percent solids. Experiments have resulted in a 23 percent-alcohol fermentation, much higher than with the conventional process. Commercial production at that level is not likely in the near future because of difficulty in staying within the required tolerances. However, incremental moves toward higher concentrations open the possibility of lower production costs.

Improved Yeast

For many years, researchers have been trying to improve yeast, which is a highly effective converter of sugars to ethanol. The desired end product is a yeast that would be more heat tolerant and better able to withstand high alcohol concentrations, that would produce fewer undesirable byproducts, and that might even be able to convert more types of sugar to ethanol. Developers have already made progress in some of these areas. For example, the ethanol tolerance of yeast is at least one-third higher today than in the 1970s.

Some researchers believe a yeast tolerant of temperatures as high as 140° F is the ideal. If such a yeast were to be developed—something increasingly possible with recombinant DNA techniques—the ethanol conversion process would look completely different than it does today (Novozymes and BBI International). Another goal of industry researchers is to produce less glycerol, which is produced in response to stress and represents a loss of ethanol during conversion.

Conversion of Pentose Sugars to Ethanol

Sucrose from starch is not the only type of sugar in the corn kernel. Some of the sugars are pentoses, or five-carbon sugars not normally utilized by common yeast. Any organism that could ferment pentoses to ethanol would be a valuable contribution to corn-ethanol conversion efficiency. This conversion has been achieved in the laboratory using genetically modified yeasts (Moniruzzaman et al.) and in bacterial fermentation using *E. coli* (Dien et al., 1997). These processes are not in commercial use, partly because the engineered organisms are less hardy and less tolerant of environmental changes than conventional organisms. Researchers are also concerned about how the nutritional content of the resulting feed coproduct would differ from conventional DDGS and about whether the genetically modified organisms remaining in the feed would be acceptable in the commercial feed market.

New Enzymes

Enzymes for Liquefaction and Saccharification

Enzymes were first used in ethanol production in the 1950s, but they have recently been improved and their cost brought down through the use of special fermentations of microorganisms. Costs have fallen 70 percent over the last 25 years (Novozymes and BBI International).

Enzymes enable chemical reactions to occur more easily, with less heat or a more moderate pH, and therefore more cost effectively. Their use in ethanol production improves liquefaction, saccharification, and fermentation. Enzyme use also results in reduced soak time, higher starch and gluten yield, better protein quality, and reduced water and energy use. USDA's Agricultural Research Service (ARS) is working with enzyme manufacturers to further reduce cost and improve effectiveness.

Enzymes to Reduce Sulfur Dioxide and Steep Time in Wet Milling

Part of the additional expense in wet milling as opposed to dry milling is the necessity of soaking the corn before separation of the germ from the kernel. The tanks increase capital cost, and the soak time slows the process. Soak time can be reduced by adding sulfur dioxide to the steep water, but research shows that the sulfur dioxide can be reduced or eliminated by using enzymes. Recently, an experimental two-stage procedure reduced soak time by up to 83 percent (Johnston and Singh; Singh and Johnston). In the saccharification step, the protease enzyme hydrolyzed the protein matrix around the starch granules and made it available for further breakdown. As with most enzymes, cost is still an issue; however, small-scale experiments seeking to optimize the process have so far reduced the enzyme requirement severalfold. Research trials show that using a low level of sulfur dioxide (more than 90 percent less than conventional levels) greatly reduces the enzyme requirement. Small amounts of sulfur dioxide are still effective in reducing bacterial contamination, a potential problem in continuous processes. Though enzymes are an added expense, the procedure has the potential to increase plant capacity (through the time savings), reduce energy costs, and allow the use of otherwise unusable broken grains. Replacing the conventional liquefaction and saccharification steps with a single, low-temperature enzyme step has already been discussed in the section "Advances in Separation Techniques."

Distillation Technology

Standard distillation techniques leave about 4-percent water in the final ethanol. In the early days of ethanol distillation, the basic production design came from the beverage alcohol industry, where there is no need to remove all the water (Swain). Fuel ethanol, however, must be almost pure or dry, so ethanol producers began dehydrating their ethanol using a technique called azeotropic distillation. This technique requires use of an ingredient, usually benzene or cyclohexane, to break the azeotrope—the point after which distillation becomes ineffective. The adoption of molecular sieves allows the modern plant to use less power, reduce original capital outlay, and eliminate potential exposure of workers to dangerous chemicals.

Molecular sieves use materials with microscopic pore sizes large enough to allow a molecule of one size to get through while blocking another. For example, in ethanol dehydration the molecular sieves have a pore diameter that allows a water molecule to enter and be trapped but keeps out the larger ethanol molecule. Since the late 1980s, vapor-phase molecular sieves have been the industry standard, improving on the liquid-phase sieves by reducing the required size (Novozymes and BBI International).

Control Systems

The 1993 prediction of cost reductions in ethanol production (Hohmann and Rendleman) overlooked the incremental changes in efficiency that were taking place due simply to increased control of fermentation and other processes. Distributed control systems were already being used, but with the evolution of technology, especially computing capabilities, these systems have continued to reduce costs while optimizing the production process.

Distributed control systems are used in industrial and other engineering applications to monitor and control a process remotely. Human operators manage equipment distributed throughout the plant (or other application). Examples, besides ethanol plants, include power distribution systems, traffic signals, water management systems, and biorefineries. Instruments to measure and control, usually digital, are wired ultimately to computers, allowing a human-to-machine interface.

Merging the distributed control system with computer programs allows timely monitoring of processes, and even allows prediction. Reports can be compiled from stored data, and alarms can be set to alert operators if established parameters are breached.

Distributed control systems have cut costs in ethanol plants mainly by reducing the labor required, but they have also improved production efficiency in other ways, letting operators fine-tune processes they could not control as closely in the past. Better process control also reduces downtime and maintenance.

Environmental Technologies

In addition to reducing the amount of energy required for production, modern ethanol plants give off fewer odors and emissions than ever before. Technological advances hold promise for converting more of the feed coproducts into ethanol, for reducing components of DDGS that might harm the environment, and for utilizing waste streams from the process.

As noted, modern ethanol production uses less energy as techniques and technologies improve. Both wet and dry millers are using less fuel and electricity per gallon produced, and farmers are producing corn more efficiently. All these savings are an environmental plus for ethanol.

Ethanol plant emissions are a second area of improvement. In 2002, Minnesota producers signed an agreement with the Environmental Protection Agency (EPA) to reduce emissions coming from their plants. That agreement has become an industry standard. Since then, ethanol producers in several States have agreed to install thermal oxidizers or other technologies that eliminate nearly all volatile organic compounds (VOCs) and other pollutants, adding equipment that averages more than $2 million per plant. The EPA estimates that the agreement will eliminate more than 63,000 tons of pollution annually.

Though pollutants, like particulate matter and even VOCs, can originate from fermentation and from other parts of the plant such as grain-handling areas, most of the attention has been focused on dryer stacks. Thermal oxidizers are now standard equipment in most new ethanol facilities. They convert carbon- and hydrogen-bearing compounds into CO_2 and water through high-temperature oxidization. Besides eliminating odors and visible emissions, thermal oxidizers can eliminate over 99 percent of oxides of nitrogen and other hazardous air pollutants, as well as certain particulates.

Wastewater emission problems have been largely solved by the development of anaerobic digester systems. Majumdar et al. report using membrane technology to recover VOCs such as hexane from process air emissions, giving membranes a role in environmental remediation.

New techniques may make processing itself more environmentally friendly. For example, a cornsteeping process being developed by scientists from ARS and the University of Illinois uses enzymes and reduces the need for sulfites.

Another discovery—one that may hold the most promise for the future—is the conversion of low-value and waste stream products into valuable coproducts. (This will be discussed in the next section, "Technologies Involving Coproducts.")

Finally, overfeeding phosphorus to animals can be an environmental concern because the phosphorus ends up on the land as manure. The phosphorus available in DDGS is more than is needed for proper animal nutrition. If the level excreted over time is excessive, phosphorus can move into ground or surface water and create problems such as algae in the waterways. Researchers are experimenting with membrane technology that would remove phosphorus from the thin stillage (the liquid remaining after removal of the wet distiller's grains) before it becomes feed. The result is likely to be more efficient feed rations and reduced environmental impact.

TECHNOLOGIES INVOLVING COPRODUCTS

In addition to animal feeds, other potential coproducts are produced along with the ethanol. Both wet and dry milling create CO_2 during fermentation. Minor components, such as glycerol, may be collected from the processing stream. Some research directions may alter the entire process, using different fermentations to produce entirely different product lines. Researchers are also hoping to turn much of the nonfuel product into ethanol, or even something more valuable than the fuel that is currently the primary product. Possibilities include protein and fiber that could be added to human foods to increase nutritional value.

Enzymatic milling may also allow recovery of valuable coproducts. Conventional dry milling leaves as coproducts distiller's dried grains with solubles and, if recovered, CO_2. ARS scientists from USDA's Agricultural Research Service are adapting the concept of enzymatic milling to the dry-mill ethanol process, partly to recover additional high-value coproducts.

The Growing Supply of Feed Coproducts

Ethanol production does not exhaust the feed value of corn; it merely uses up the starch portion, leaving protein, minerals, fat, and fiber to be dried and sold for feed. A high percentage of ethanol producers' revenue comes from the feed coproduct. Dry milling turns a bushel of corn (56 lb) into 2.7 or more gallons of ethanol and leaves 17 lb of distiller's grains. Wet milling produces only slightly less ethanol and coproduces around 16 lb of corn gluten meal (CGM, 2.65 lb) and corn gluten feed (CGF, 13.5 lb), in addition to corn oil. In 2004, 7.3 million metric tons of DDGS, 426,400 metric tons of CGM, and 2.36 million metric tons of CGF were produced (Renewable Fuels Association). Predictions for near-term increases are as high as 10 million tons by 2007-08, which would constitute a significant portion of all the cattlefeed in the United States (*BioFuels Journal*, 2005c).

Because feed coproducts and ethanol are produced in fixed proportions, increased demand for ethanol will result in greater output of DDGS, putting downward pressure on its price. Partly for this reason, research is continuing on alternative coproducts.

Proving the nutritional value of DDGS for new uses will expand the market. Most of the DDGS produced is fed to dairy and beef cows, but it is increasingly being tested and used in swine and poultry rations. Research is planned on DDGS in equine diets.

The nonuniform character of DDGS makes it difficult to establish feeding parameters because the product varies in consistency and nutritional value. Ongoing research is aimed at establishing feed values for various forms of DDGS, and some producers are developing proprietary DDGS brands with guaranteed nutritional properties.

Sequential Extraction

Lawrence Johnson of Iowa State University has initiated several projects exploring the parameters of the sequential extraction process, or SEP (see, for example, Hojilla-Evangelista and Johnson). SEP uses alcohol rather than water to separate kernel components in an otherwise wet-mill process. The researchers claim increased ethanol production (10 percent), higher quality protein extracts (with no SO_2 needed for extraction, and therefore less degradation of the protein portion), and the production of corn fiber gum, a gum arabic substitute, as a coproduct. Gum arabic is used in producing soft drinks, candy, and pharmaceuticals. Cost remains a problem with the SEP process: The initial capital outlay for an SEP plant would be much higher than for a conventional wet mill. Extraction of proteins from the separated germ remains a problem as well.

Corn Germ Recovery for the Dry-Mill Process

As explained in the "Process Improvements" section, both dry fractionation and the quick-germ technique modify the dry-mill process and make the recovery of corn germ possible. Recovery of the germ, and thus of the oil, can make dry milling more profitable. The soak step in the quick-germ process takes less time than in wet milling, 3-6 hours as opposed to 24-48 hours. The process has not yet been used commercially. Plants using dry fractionation, however, are underway. Profitability will depend on the relative cost of corn oil and the capital costs of the additional equipment required.

Centrifugal Corn Oil Separation from the Distiller's Grain Stream

An additional technology for separating corn oil from distiller's grain has recently become available to dry millers. SunSource, a coalition of ethanol producers and a technology company, is licensing a system that can be installed in new plants or retrofitted to existing ones (Walker). The technology makes an additional coproduct available to ethanol producers, most likely to be used in biodiesel production. The process uses centrifuge technology to extract the oil from the distiller's grains in the evaporation step. The developers of the process claim that removing the oil from the distiller's grains does not lower the value of the feed coproduct and makes it easier to handle. The process also reduces volatile organic compounds emitted from the dryers, an environmental bonus.

CO_2 Recovery

Ethanol's most abundant coproduct is CO_2, produced by yeast in about the same proportion as ethanol itself. Only about 25 U.S. plants find uses for the gas (Lynn Grooms in *BioFuels Journal,* 2005b); the other plants, because of the low commercial value of CO_2, simply vent it into the air. Most CO_2 sold commercially is used in soft drinks and food processing, while other uses, such as water treatment, welding, chemical processing, refrigerants, and hydroponics, consume some of the remainder.

Another approach would be to find new uses for the gas, raising its value and expanding the market. One experiment uses CO_2 to enhance the recovery of oil from depleted oilfields. The gas is pumped into the oil production zones, forcing residual oil to the surface. If successful, the technique could greatly expand the market for CO_2.

A number of experiments with CO_2 are taking place at the basic science level (Bothast). One idea, successful at that level, is to turn the gas into ethanol or other fuel. However, the techniques are not yet commercially viable.

Bioconversion offers hope for increased CO_2 exploitation. Through biological processes, it turns organic materials into usable products or energy. In corn ethanol production, the possibility of bacterial bioconversion of CO_2 into fuel (e.g., ethanol or methanol) is under study.

Stillage Clarification and Other Uses of Membranes

Membrane separation is now used in dry mills to treat incoming boiler water and in wet mills to clarify dextrose. Based on the molecular size of the particles permitted to pass through, membranes are classified as reverse osmosis, nanofiltration, ultrafiltration, or microfiltration. They are made of various materials, including organic polymer, ceramics, and stainless steel.

Membranes were once thought promising for removing the last of the water from ethanol—the portion left by ordinary distillation—as processors looked for substitutes for hazardous materials like benzene that were then used for that purpose. In plants built today, however, the drying part of ethanol production is done with molecular sieves.

Membrane systems are used in industry to separate, clarify, and concentrate various feed streams and are most efficient when used on dilute broths. In the future, they may be used to reduce the cost of production through recovering and purifying minor components of the ethanol product stream. The recovery of lactic acid and glycerol from thin stillage is an application of this research that may soon be commercially viable. Corn oil and zein, a protein, are also potentially recoverable (Kwiatkowski and Cheryan). Researchers think membranes can be useful in continuous (though not in batch) reactors (Escobar et al.).

Biorefinery

One definition for a biorefinery is "a facility that integrates biomass conversion processes and equipment to produce fuels, power, and value-added chemicals from biomass." The concept is a bit like wet milling, but with more coproduct possibilities. Corn or another feedstock might be decomposed and recombined into a number of products other than, or in addition to, ethanol. Some of the products might be high-value, low-volume ones, outside the traditional ethanol market, that could supplement a plant's primary, lower value product line.

Extraction of Compounds from DDGS

The biorefinery concept meshes with a new area of research known as functional foods or nutraceuticals. Two categories of functional foods, dietary fibers and oligosaccharides, can be derived from grains.

Several Canadian companies and some universities are experimenting with extracting new coproducts from the distiller's grains generated from alternative feedstocks. Examples include cosmetics from oat derivatives, phenolic avenanthramides (useful in low-density liproprotein resistance to oxidation) from oats and wheat, and beta-glucan, a fiber-type complex sugar derived from yeast, oats, and barley fiber and useful in reducing cholesterol.

A biorefinery may also yield purified proteins used as human food or in industrial processes. A potentially recoverable coproduct is zein, a corn protein that can be used in adhesives and in coatings for pharmaceuticals and packaging materials because of its good water-vapor barrier properties. ARS scientists are working to develop a cost-effective way to recover zein from the milling stream, potentially making its use more common.

Corn Fiber Oil Recovery

Corn fiber is a byproduct of corn wet milling and may be a future product of dry milling. ARS has been developing methods to obtain enriched protein, starch, and fat from corn fiber. Corn fiber is also a source of corn fiber oil, a valuable dietary supplement that contains high levels of cholesterol-lowering and antioxidant phytosterols. Corn fiber can also be recovered by other means, such as the quick-fiber technique. Its economical recovery could give ethanol production another valuable coproduct line.

REGIONAL IMPACTS OF ETHANOL PLANTS

Ethanol production holds promise for rural communities that hope to add value to locally produced corn. A new ethanol plant is seen as a way to create jobs and revitalize the local rural economy. Such scenarios seem more likely with widely dispersed, smaller scale plants—the kind that could spring up to service small specialty markets with an unusual coproduct or to take advantage of a local feedstock or other regional characteristic.

In the future, corn-processing facilities may be able to recover special healthful elements. One example might be proteins recovered from the process stream by ultrafiltration and made available as food.

Minor producers may take advantage of discard items like soft drinks or candy past their expiration date or other unusual sources of feedstock. Two cheese plants in California produce ethanol and at the same time solve a disposal problem by using the whey—generally regarded as waste—as a feedstock. Cheese whey contains lactose that, if not fermented, may require special processing at the cheese plant to avoid extra charges for municipal water-disposal treatment.

Some ethanol producers take advantage of their location to reduce production costs by selling coproducts otherwise too bulky to transport, such as steep-water. Or they save by avoiding drying, for instance, by feeding their wet distiller's grains to cattle or even fish. Others may sell CO_2, or may at least capture some value from it by using it in greenhouses to boost plant growth.

With the development of commercial cellulosic techniques, ethanol producers may actually be paid to take away material that they can use for feedstock—for instance, waste from other production systems or even municipal garbage.

Other potential niche markets may use animal waste. One current USDA research project seeks to exploit the advantages of co-location (of an ethanol plant and poultry farms) by generating electric power and steam for ethanol production from chicken litter. Coproduction of power and steam from a waste stream is being developed in another USDA project.

Many ethanol facilities recently opened or under construction are farmer owned and dispersed across rural areas. Farmers expect the completion of an ethanol plant in the area to increase the local corn price. McNew and Griffith examined the impact on local corn prices of opening 12 ethanol plants and found that, on average, price per bushel rose 12.5 cents at the plant and that some price response was detected 68 miles from the plant.[1]

New rural ethanol plants also provide employment. Though the plants need fewer employees than they would have just a few years ago when the industry was using less labor-saving technology, the U.S. Department of Energy (DOE) estimates that ethanol production is responsible for 40,000 jobs and $1.3 billion in increased annual household income.

National Benefits from Ethanol

Ethanol produced in the United States displaces imported foreign oil and creates domestic economic activity. Gallagher et al. (2000) estimate that the current program results in a $400-million net gain in overall social welfare. More domestic production, which is likely with an MTBE phaseout, would result in additional gain.

With ethanol production of 3.41 billion gallons in 2004, 143.3 million fewer barrels of oil were needed, about 4.5 percent of annual U.S. use (Urbanchuk). A 10-percent mixture of E-10 reduces petroleum use by 6 percent, greenhouse gas emissions by 1 percent, and fossil energy use by 3 percent (Wang et al.).

Increased biomass production would also change the picture. Gallagher and Johnson focus primarily on the benefits of a developed biomass-to-ethanol industry. Assuming an industry based on corn stover, the largest single source of biomass, they conclude that U.S. welfare would increase (a) because of the expanded fuel supply and (b) because the oligopoly effects of pricing by the Organization of Petroleum Exporting Countries (OPEC) would be mitigated. A biomass-to-ethanol industry, based on something like switchgrass, could add to farmers' product line and "could significantly increase profits for the agricultural sector" (De La Torre Ugarte et al.).

Biomass: Ethanol's Future?

Though corn has been the feedstock of choice in the United States, ethanol potentially can be made from any starch, sugar, or cellulosic feedstock. In fact, ethanol has been created from a variety of grains and from grass and straw, wood fibers, and sugarcane. Though ethanol production from corn has become more efficient, some experts see it as a technology that has already matured, with any significant reduction in production costs unlikely (DiPardo). Substantial cost reductions may be possible, experts believe, if cellulose-based feedstocks are used instead of corn. One industry publication editorializes: "Ultimately, if renewable automotive fuel becomes economical in the United States it will have to be made from lignocellulosic biomass" (*Industrial Bioprocessing*).

In the end, biomass-to-ethanol production may be attractive because biomass would cost less than corn. In addition, selling the feed coproducts from corn ethanol may become burdensome. As corn ethanol production increases, an inevitable result of dry-mill expansions, more feed coproducts will find their way into the market and drive down prices for DDGS.

The vision of cellulosic conversion is not yet commercial reality, however, due to difficulties inherent in turning biomass into ethanol. Because the cellulose and fermentable portions of woody biomass are tightly bound together, researchers have had to focus on the problem of pretreatment and hydrolysis. The necessary chemical conversion can take place using acids or enzymes, but, to be commercially viable, costs must be brought down. USDA and the U.S. Department of Energy are funding projects that have this aim. Also, many of the sugars making up cellulosic feedstocks (composed of cellulose and hemicellulose) are not readily convertible to ethanol by ordinary yeast.

Researchers have not settled on the cost of producing ethanol from biomass because ethanol is not yet being produced this way. A joint project by the USDA and DOE estimated the unit cost at a 25-million-gallon per year (MMgy) plant to be $1.50 per gallon (1999 dollars) (McAloon et al.). Though some researchers put this figure lower, at $1.16-$1.44 per gallon (Wooley et al.), recent estimates of the cost of commercial cellulose collection ($40-$50 per ton) and capital outlays (over $6 per gallon of annual capacity) make the likely cost higher. A consideration for new facilities will be the one-of-a-kind expense associated with setting up the first working biomass-to-ethanol plants (e.g., scaleup and development costs). In its $1.50-per-gallon estimate, the USDA/DOE study assumed these costs had already been incurred.

Cellulose to Ethanol: The Process

In the same way that starch from corn must be hydrolyzed and saccharified (decomposed further into simple sugars) before it can be fermented, cellulose must first be converted to sugars before it is fermented and turned into ethanol. Cellulosic feedstocks are more difficult than corn to convert to sugar. Cellulose can be converted by dilute acid hydrolysis or concentrated acid hydrolysis, both of which use sulfuric acid. Hydrolysis can also be achieved with enzymes or by other new techniques, including countercurrent hydrolysis or gassification fermentation.

Cellulosic hydrolysis produces glucose and other six-carbon sugars (hexoses) from the cellulose and five-carbon sugars (pentoses) from hemicellulose. The non-glucose (carbon) sugars must be fermented to produce ethanol, but are not readily fermentable by *Saccharomyces cerevisia*, a naturally occurring yeast. However, they can be converted to ethanol by genetically engineered yeasts, though the process is not yet economically viable (DiPardo).

Supplying Biomass

Some research has centered on converting the cellulosic portion of the corn kernel. Many of the researchers in this effort initially believed that cellulosic conversion could begin with the cellulose in the current feedstock stream (corn) and proceed to corn husks, and then to corn stover, before they finally extended their research to include other sources of cellulose. A second line of research has centered on converting cellulosic biomass directly into ethanol from noncorn sources such as small-diameter trees and switchgrass. Further research has focused on converting cellulose after it has entered the waste or coproduct stream, through steam explosion, for example.

If converting biomass to ethanol can be made economically attractive, the potential feedstocks are myriad. They include agricultural waste, municipal solid waste, food processing waste, and woody biomass from small-diameter trees. Agronomic research is underway on improving dedicated-energy crops such as hybrid willow, hybrid poplar, and switchgrass (DiPardo). One project looks at genetic improvement of switchgrass to optimize its conversion. Grasses grow quickly, of course, while tree crops such as willow require a 22-year rotation, with the first harvest in year 4 and subsequent harvests every 3 years thereafter. Hybrid poplar trees require 6-10 years to reach their first harvest.

Biomass from crop residue can be a source of farm profits, and with appropriate steps, its use in ethanol conversion can be environmentally friendly. A USDA study by Gallagher et al. (2003) concluded that crop residues are the cheapest prospective source of fuel for the U.S. market, with the energy potential to displace 12.5 percent of petroleum imports or 5 percent of electricity consumption.

Biomass Byproducts: Problems with Acid and High Temperatures

The dilute acid and concentrated acid hydrolysis used in biomass-to-ethanol conversion produce byproducts that either must be disposed of or that require recycling of sulfuric acid. In addition, the high temperatures required take a toll on the sugars and thus on the ethanol yield (DiPardo).

Potential Solutions to Acid and Heat Problems

Countercurrent Hydrolysis—Advances in biotechnology could reduce conversion costs substantially. The National Renewable Energy Laboratory (NREL) presents countercurrent hydrolysis as a new pretreatment. The process uses steam to hydrolyze most of the hemicellulose. In a second, hotter, stage, dilute sulfuric acid hydrolyzes the rest of the hemicellulose and most of the cellulose. NREL researchers believe the countercurrent hydrolysis process offers more potential for reducing costs than the dilute sulfuric acid process. They estimate that it will increase glucose yields and permit a higher fermentation temperature, resulting in an increased yield of ethanol. They have achieved glucose yields of over 90 percent in experiments with hardwoods (NREL, 2004).

A New Approach—While acknowledging that a cellulose-to-ethanol industry is in its infancy, Tembo et al. attempt to determine parameters that would define a regional (Oklahoma) biorefinery industry. The authors assess alternatives for future production, positing a gassification-fermentation technique that uses neither traditional acid nor enzymatic hydrolysis. Lignocellulosic biomass can be gassified, they say, in fluidized beds to produce "synthesis gas." The gas can then be converted by anaerobic bacteria to ethanol. Advantages to such a lignocellulosic system include a theoretically lower cost than for corn ethanol and the potential to use multiple perennial feedstocks, with supposedly less environmental impact than corn use. The researchers caution that the process is at the bench level and is not yet commercially available.

Enzymatic Hydrolysis—NREL believes that the greatest potential for ethanol production from biomass lies in enzymatic hydrolysis of cellulose. Advances in biotechnology may eventually make the technique possible through the use of genetically engineered bacteria and may also permit the fermentation of the pentoses.

Reducing Enzyme Costs

The use of enzymes in biomass-to-ethanol conversion will require reductions in the cost of producing cellulase enzymes, along with an increased yield in the conversion of nonglucose sugars to ethanol. The enzyme cellulase, already used in industry, replaces sulfuric acid in the hydrolysis step. Higher sugar yields are possible because the cellulase can be used at lower temperatures (Cooper). NREL reports that recent process improvements allow simultaneous saccharification and fermentation, with cellulase and fermenting yeasts working together so that sugars are fermented as they are produced. NREL estimates that cost reductions could be four times greater for the enzyme process than for the concentrated acid process and three times greater than for the dilute acid process.

In speaking of both corn stover and forest product waste, *Industrial Bioprocessing* writes, "The big stumbling block in manufacturing ethanol from biomass is the cost of hydrolyzing cellulose into fermentable sugar." To speed the quest for cheaper ways of using enzymes to convert biomass to ethanol, DOE funded research by two commercial enzyme companies, Novozymes and Genencor International Inc. Cellulase enzymes were recently reported to cost 45 cents per gallon of finished ethanol, making them too expensive for commercial use (NREL, 1998). NREL estimated that the cost could be reduced to less than 10 cents with scaled-up production. In fact, Novozyme recently announced that its researchers have successfully completed this project and have reduced the enzyme costs to 10-18 cents per gallon of ethanol produced (Susan Reidy in *BioFuels Journal,* 2005a).

Other Biomass-to-Ethanol Improvements

NREL estimates that, in addition to reduced costs of enzyme conversion, improvements in acid recovery and sugar yield for the concentrated acid process could save 4 cents per gallon and that process improvements for the dilute acid technology could save about 19 cents per gallon.

Considerable success in cellulosic conversion has already been achieved at the experimental level. Depending on the prices of alternatives and the success of current scaleup efforts, commercial viability may be possible in the near term. A Canadian firm, Iogen Corporation, in partnership with the Canadian Government, has demonstrated a process that turns wheat straw into fermentable sugar.

CONCLUSIONS: ETHANOL'S POTENTIAL

In keeping with USDA's early estimates of the savings to come, the cost of ethanol production has indeed fallen. Though improvements in process optimization and technology have been important, the fluctuating prices of inputs such as corn, the price of energy alternatives, and even environmental developments such as a drop in MTBE use, play larger roles in the fortunes of the industry.

Ethanol production is becoming a mature industry, with savings in the next 10 years likely to be smaller than those of the last 10-15 years.

Some developments, such as dry fractionation technology—soon to be commercially employed—may alter the structure of the industry by giving an edge to the less capital-intensive dry-mill method. This advantage for dry milling may make it easier to build smaller plants that are cost competitive, and local farmer cooperatives could flourish as a result.

Promising areas to be exploited by new technology include coproduct development, cellulosic conversion, and niche markets. The recovery of high-value food supplements may reduce financial risk by giving the industry an outlet outside the capricious energy market. Continued high oil prices may spur the risk-taking necessary to overcome the initial scaleup and development costs of cellulose-to-ethanol production. Niche markets that take advantage of locally available feedstocks, that have local outlets for coproducts, or that produce unique coproducts may also contribute to the industry's growth.

REFERENCES

BioFuels Journal. (2003a). "Analyze This: Online Tool Shows Impact of Ethanol Plant on Area Corn Prices," Fourth Quarter, p.16. McNew and Griffith's online tool, accessed at http://extensionecon.msu.montana.edu/eplantanalyzer/

BioFuels Journal (2003b). "Selecting Hybrids for Dry-Mill Ethanol," Fourth Quarter, pp. 18-19.

BioFuels Journal. (2005a). "Biomass-to-Ethanol for Less: NREL, Novozymes Announce 30-Fold Reduction in Enzyme Cost," Second Quarter, pp. 107-8.

BioFuels Journal. (2005b) "Sparking CO_2 Interest: KS Oil Field Pilot Project Could Open New Market for Ethanol Plants," Second Quarter, pp. 8-9.

BioFuels Journal. "The 10M Ton DDG Question: Speakers Explore Options for Use of Distiller's Grains, (2005c)." Second Quarter, pp. 86-87.

BioFuels Journal. (2005d). "Fractionating Energy: Corn Fractionation Processes Help Cut Energy Costs," Third Quarter, pp. 31-32.

BioFuels Journal. (2005e). "FWS Dry Fractionation Processes," Third Quarter, p. 32.

Bothast, Rodney. Personal interview, September 9.

Bryan, Tom. (2004). "A Long View Approach," *Ethanol Producer Magazine,* June, p. 42.

Bryan, Tom. (2002) "The Search for a Perfect Ethanol Corn," *Ethanol Producer Magazine,* September, pp. 14-19.

Canadian Renewable Fuels Association, (1999) accessed at http://www.greenfuels.org/ethaques.html

Cooper, C. "A Renewed Boost for Ethanol," *Chemical Engineering 106(2)*, February.

"De La (2003) Torre Ugarte, G. Daniel, Marie E. Walsh, Hosein Shapouri, and Stephen P. Slinsky. *The Economic Impacts of Bioenergy Crop Production on U.S. Agriculture,* AER-816. USDA, Office of Energy Policy and New Uses, February.

Dien, B.S., Bothast, R.J. Iten, L.B. Barrios, L. & Eckhoff, S.R. (2002). "Fate of Bt Protein and Influence of Corn Hybrid on Ethanol Production," *Cereal Chemistry 79(4)*:582-85, March.

Dien, Bruce S., (2002). Rodney J. Bothast, Nancy N. Nichols, and Michael A. Cotta. "The U.S. Corn Ethanol Industry: An Overview of Current Technology and Future Prospects," *International Sugar Journal 104(1241)*:204-11, April.

Dien, B.S., Hespell, R.B. Ingram, L.O. & Bothast, R.J. (1997). "Conversion of Corn Milling Fibrous Coproducts Into Ethanol by Recombinant *Escherichia coli* Strains K011 and SL40," *World Journal of Microbiology and Biotechnology (13)6:*619-25, November.

DiPardo, Joseph. (2002). "Outlook for Biomass Ethanol Production and Demand." Page last modified July 30. Accessed at http://www.eia.doe.gov/oiaf/analysispaper/biomass.html, February 20, 2004.

Eckhoff, Steve. (2001). "Use the Quick Germ Method—Lower Your Ethanol Production Costs," *Seed World 139(3)*:14-16, March.

Eckhoff, S.R., Du, L. Yang, P. Rausch, K.D. Wang, D.L. B.H. & Lin, M.E. (1999). Tumbleson. "Comparison Between Alkali and Conventional Corn Wet-Milling: 100-g Procedure," *Cereal Chemistry 76(1):*96-99.

Escobar, J.M., Rane, K.D. & Cheryan, M. (2001). "Ethanol Production in a Membrane Bioreactor: Pilot-scale Trials in a Corn Wet Mill," *Applied Biochemistry and Biotechnology: 22nd Symposium on Biotechnology for Fuels and Chemicals 91-93 (1-9):*283-96, Spring.

Feed Situation and Outlook Yearbook. (2003). USDA, Economic Research Service, various years.

Gallagher, Paul, Mark Dikeman, John Fritz, Eric Wailes, Wayne Gauther, and Hosein Shapouri. *Biomass From Crop Residues: Cost and Supply Estimates,* AER-619. USDA, Office of Energy Policy and New Uses, March.

Gallagher, Paul, and Donald Johnson. (1999). "Some New Ethanol Technology: Cost Competition and Adoption Effects in the Petroleum Market," *Energy Journal 20(1999):*89-120.

Gallagher, Paul W., Daniel M. (2000). Otto, and Mark Dikeman. "Effects of an Oxygen Requirement for Fuel in Midwest Ethanol Markets and Local Economies," *Review of Agricultural Economics, 22(2)*:292-311.

Haefele, D., Owens, F. O'Bryan, K. & Sevenich, D. *Selection and Optimization of Corn Hybrids for Fuel Ethanol Production.* In Proceedings, ASTA 59th Annual Corn and Sorghum Research Conference, 2004. CD-ROM. Alexandria, VA, American Seed Trade Association.

Hohmann, Neil, & Matthew Rendleman., C. (1993). *Emerging Technologies in Ethanol Production.* AIB-663, USDA, Economic Research Service, January.

Hojilla-Evangelista, Mila P., & Lawrence A. Johnson. "Sequential Extraction Processing of High-Oil Corn," *Cereal Chemistry 80(6):*679-83, November-December 2003.

Illinois Corn Growers Association and the Illinois Corn Marketing Board, accessed at http://www.ilcorn.org/Pages/fstintro.htm, September 14, 2000.

Industrial Bioprocessing, (2003). *"Enzymes for Biomass Ethanol," 25(4),* April.

Johnston, David B., & Vijay Singh. (2001). "Use of Proteases to Reduce Steep Time and SO(2) Requirements in a Wet-Milling Process," *Cereal Chemistry 78(4)*:405-11.

Krohn, Bradley. (2003). Former technical leader of Monsanto's High Fermentable Corn Program, personal interview, September 9,.

Kwiatkowski, Jason R., & Munir Cheryan. "Recovery of Corn Oil From Ethanol Extracts of Ground Corn Using Membrane Technology," *Journal of the American Oil Chemists' Society 82(3):*221-27, 2005.

Levelton Engineering Ltd. and (S&T)2 Consulting Inc. "What Is Driving the Fuel Ethanol Industry?" available at http://www/jgpc.ca.

Madlinger, Christy. (2005). *Director of Technical Services*, Frazier, Barnes & Associates, personal interview, September 15.

Majumdar, S., Bhaumik, D. Sirkar, K.K. & Simes. G. (2001). "A Pilot-Scale Demonstration of a Membrane-Based Absorption-Stripping Process for Removal and Recovery of Volatile Organic Compounds," *Environmental Progress 20(1)*:27-35.

McAloon, Andrew, Frank Taylor, Winnie Yee, Kelly Ibsen, and Robert Wooley. (2000). *Determining the Cost of Producing Ethanol From Corn Starch and Lignocellulosic Feedstocks*. Technical Report of National Renewable Energy Laboratory (NREL), Golden, CO, October.

McNew, Kevin, & Duane Griffith. (2005). "Measuring the Impact of Ethanol Plants on Local Grain Prices," *Review of Agricultural Economics 27(2)*:164-80, June.

Moniruzzaman, M., B.S. Dien, C.D. Skory, Z.D. Chen, R.B. Hespell, N.W.Y. Ho, Dale, B.E. & Bothast, R.J. (1997). "Fermentation of Corn Fibre Sugars by an Engineered Xylose Utilizing Saccharomyces Yeast Strain," *World Journal of Microbiology & Biotechnology (13)*:341-46,.

National Renewable Energy Laboratory (NREL). (2004). "Biofuels – Pretreatment Technology," accessed at http://www.ott.doe.gov/biofuels/current_research.html, April 15.

National Renewable Energy Laboratory (NREL), (1998). "The Cost of Cellulase Enzymes," in *Bioethanol From the Corn Industry*, DOE/GO-10097-577, Golden, CO, May.

Novozymes and BBI International. (2004). *Fuel Ethanol: A Technological Evolution*. Grand Forks, ND: BBI Publishing, June.

Rendleman, C. Matthew, and Neil Hohmann, "The Impact of Production Innovations in the Fuel Ethanol Industry," *Agribusiness, 9(3)*:217-31, 1993.

Renewable Fuels Association. (2005). *Homegrown for the Homeland: Ethanol Industry Outlook 2005*. Washington, DC, February.

Ross, Martin. "California Ethanol Dream Coming to Fruition," *FarmWeek*, April 21, 2003, p. 5.

Rutherford, Amy, Director of Monsanto's "Fuel Your Profits" program, interview August 5, (2004).

Shapouri H., Duffield, J.A. & Wang, M. (2002). *The Energy Balance of Corn Ethanol: An Update*, AER-813. USDA, Office of Energy Policy and New Uses, July.

Shapouri, Hosein, & Paul Gallagher. (2005). *USDA's 2002 Ethanol Cost of Production Survey*, AER-841. USDA, Office of Energy Policy and New Uses, July.

Singh,V., & S.R. Eckhoff.(1997) "Economics of Germ Preparation for Dry-Grind Ethanol Facilities," *Cereal Chemistry 74(4)*:462-66.

Singh, V., & Eckhoff, S.R.(1996) "Effect of Soak Time, Soak Temperature, and Lactic Acid on Germ Recovery Parameters," *Cereal Chemistry 73(6)*:716-20.

Singh, V., & Graeber, J.V. (2005). "Effect of Corn Hybrid Variability and Planting Location on Dry Grind Ethanol Production." *Transactions of the American Society of Agricultural Engineers 48(2)*:709-14.

Singh, Vijay, & David B. Johnston. (2004). "Enzymatic Corn Wet Milling Process: An Update," paper presented at the 2004 Corn Utilization and Technology Conference, Indianapolis, IN, June 7-9.

Swain, R.L. Bibb. (1999). "Molecular Sieve Dehydrators: How They Became the Industry Standard and How They Work." Chapter 19 in *The Alcohol Textbook*, 3rd ed., pp. 289-93, Nottingham, England: University Press.

Taylor, Frank, James C. Craig, Jr., M.J. Kurantz, and Vijay Singh. (2003). "Corn-Milling Pretreatment with Anhydrous Ammonia," *Applied Biochemistry and Biotechnology 104(2):*41-48, February.

Taylor, Frank, Andrew J. McAloon, James C. Craig, Jr., Pin Yang, Jenny Wahjudi, & Steven R. Eckhoff. (2001). "Fermentation and Costs of Fuel Ethanol from Corn with Quick-Germ Process," *Applied Biochemistry and Biotechnology 94(1):*41-50, April.

Tembo, Gelson, Francis M. Epplin, and Raymond Huhnke. (2003). "Integrative Investment Appraisal of a Lignocellulosic Biomass-to-Ethanol Industry," *Journal of Agricultural and Resource Economics 28(3):*611-33, December.

Urbanchuk, John M. (2005) "Contribution of the Ethanol Industry to the Economy of the United States," paper prepared for the Renewable Fuels Association, January 10.

Walker, Devona. (2005). "New Technology May Converge Biodiesel, Ethanol Markets." Associated Press, June 28, 2005, accessed at http://www.aberdeennews.com. September 19.

Wang, M.Q. (1999). *GREET 1.5 – Transportation Fuel-Cycle Model: Volume 1: Methodology, Development, Use, and Results*, Center for Transportation Research, Argonne National Laboratory, Argonne, IL, August.

Wang, M., C. Saricks, & Santini, D. (1999). *Effects of Fuel Ethanol Use on Fuel-Cycle Energy and Greenhouse Gas Emissions*. Center for Transportation Research, Argonne National Laboratory, Argonne, IL, January.

Wooley, Robert, Mark Ruth, David Glassner, & John Sheehan. (1999). "Process Design and Costing of Bioethanol Technology: A Tool for Determining the Status and Direction of Research and Development," *Biotechnology Progress 15(5):*794–803, September-October.

End Notes

[1] McNew and Griffith have also established a Web site at Montana State University that shows how corn prices are impacted by the opening of an ethanol plant (*BioFuels Journal*, 2003a). The site allows the user to investigate a small, medium, or large plant and displays the results graphically in map format. The impacts are based on econometric work by the authors. The tool, called the "Ethanol Plant Analyzer: A GIS-Driven Tool for Assessing Ethanol Feasibility," can be found at http://extensionecon. msu.montana.edu/eplantanalyzer/

In: Ethanol Biofuel Production
Editor: Bratt P. Haas

ISBN: 978-1-60876-086-2

Chapter 2

RESEARCH ADVANCES - CELLULOSIC ETHANOL

National Renewable Energy Laboratory

FUELING THE FUTURE

On the Road to Energy Security

One of our greatest challenges is to reduce our nation's dependence on imported petroleum. To accomplish this, we need a variety of alternative fuels, including ethanol produced from cellulosic materials like grasses and wood chips. Fortunately, the United States has abundant agricultural and forest resources that can be converted into biofuels. Recent studies by the U.S. Department of Energy (DOE) suggest these resources can be used to produce enough ethanol – 60 billion gallons/year – to displace about 30% of our current gasoline consumption by 2030.

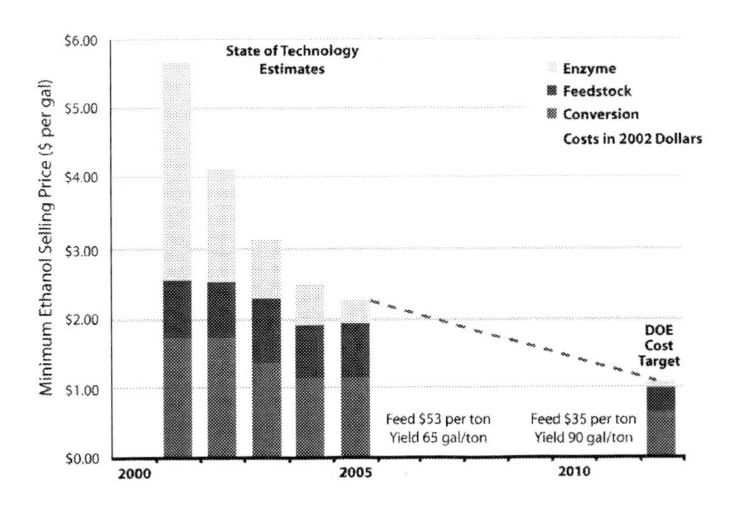

NREL has led progress toward DOE's cost target of $1.07/gallon for biochemically produced ethanol. Similar progress is being made for thermochemically produced ethanol.

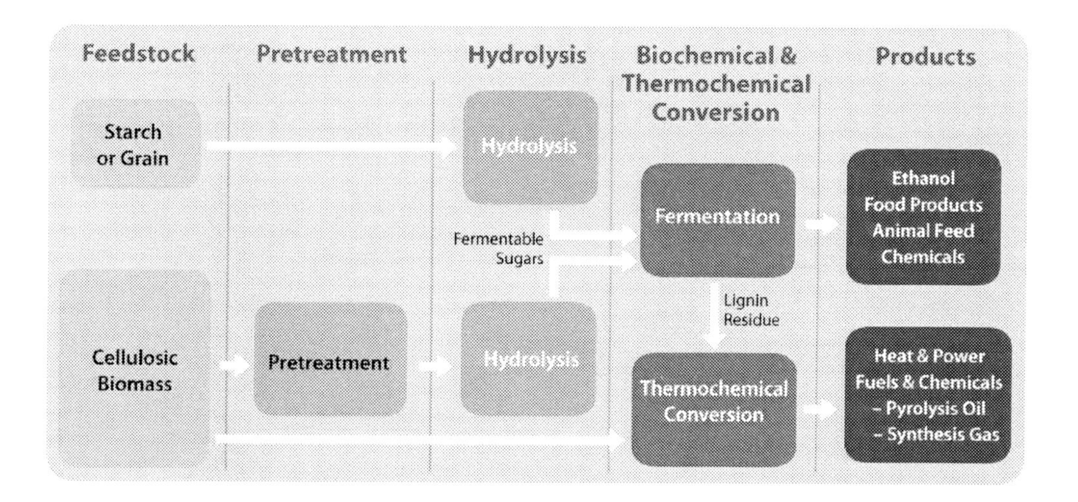

Schematic of a hypothetical integrated ethanol biorefinery.

How do we get there? Currently, there are no commercial cellulosic ethanol refineries. The ethanol we use is derived primarily from corn kernels, a form of starchy biomass. When manufacturers produce ethanol from corn, they use enzymes to convert starches to simple sugars and yeasts to ferment the sugars into ethanol. Cellulosic biomass contains sugars as well, but they are much harder to release than those in starchy biomass. To complicate matters, the process of releasing the sugars produces by-products that inhibit fermentation, and some of the sugars from cellulosic biomass are difficult to ferment.

All this makes cellulosic ethanol production complicated—and expensive. To displace petroleum, cellulosic ethanol must be cost competitive. DOE has determined that competitiveness can be achieved at an ethanol production cost of $1.07/gallon (in 2002 dollars) and aims to achieve this goal by 2012. To do this, the technology used to produce cellulosic ethanol must be improved. That's where the National Renewable Energy Laboratory (NREL) comes in.

NREL leads DOE's National Bioenergy Center and is on the cutting edge of cellulosic ethanol technology. NREL's research addresses each step of the processes that produce cellulosic ethanol and valuable co-products. NREL's research covers the full spectrum from fundamental science and discovery to demonstration in fully integrated pilot plants.

This brochure highlights some of NREL's recent advances in cellulosic ethanol production. Research at NREL addresses both biochemical (chemicals, enzymes, and fermentative microorganisms) and thermochemical (heat and chemical) processes. For the biochemical processes, NREL investigates pretreatment, hydrolysis, and fermentation steps as well as process integration and biomass analysis. For the thermochemical processes, NREL researches catalyst development, process development, and process analysis.

PRETREATMENT

Improving the Critical First Step toward Cost-Competitive Ethanol

To break down cellulose—the primary source of sugar in fibrous biomass—you have to first get past hemicellulose and lignin, which surround the cellulose in a protective sheath. This is the job of pretreatment. NREL typically uses a moderately high-temperature, high-pressure dilute acid pretreatment process to break down (hydrolyze) hemicellulose and disrupt or dissolve lignin. Hydrolyzing the hemicellulose also creates another important source of soluble sugars for later fermentation to ethanol.

NREL is investigating potentially cheaper, but still effective, pretreatment methods. In one recent advance, NREL researchers applied their knowledge of biomass structural changes to pretreatment process development. Lignin dissolved under certain pretreatment conditions can apparently redeposit onto cellulose, creating a barrier to effective cellulose hydrolysis and reducing sugar yield. NREL is using its state-of-the-art imaging and analytical tools to understand lignin redeposition and design pretreatment processes that minimize its detrimental effects.

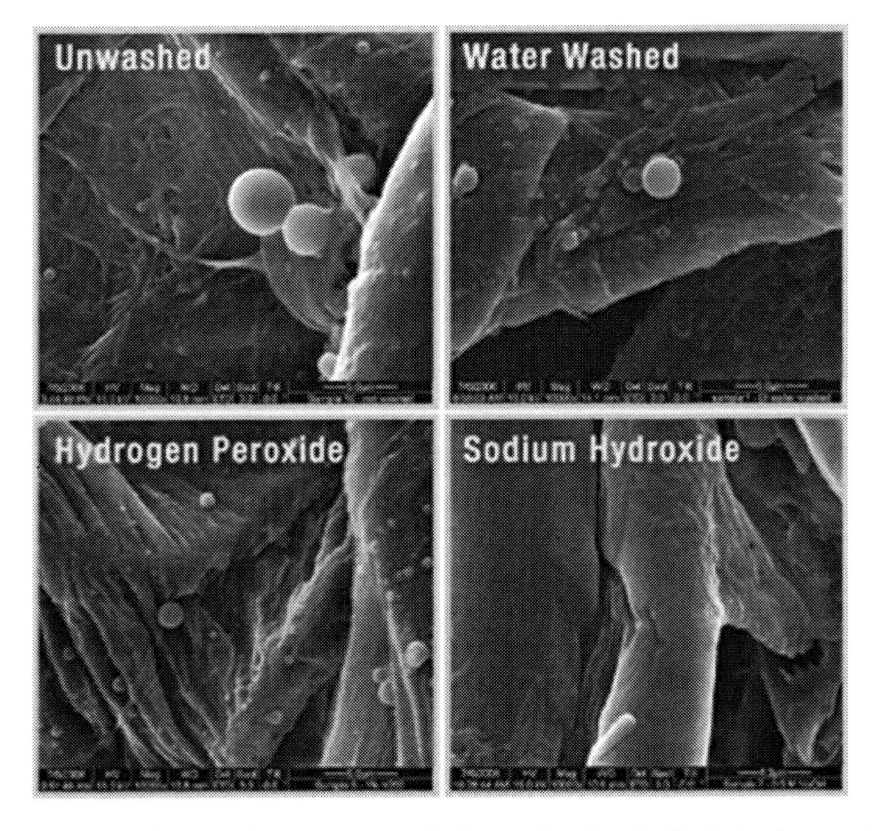

NREL used scanning electron microscopy to reveal what are thought to be lignin droplets remaining on pretreated filter paper after washing with various solvents. (T. Vinzant, NREL)

In another recent advance, NREL employed enzymes to enable milder pretreatment. Although dilute acid pretreatment can break down hemicellulose very effectively, the severe conditions require expensive processing equipment and tend to degrade the sugars. Using a milder pretreatment process could cut process costs dramatically and eliminate sugar degradation losses. The challenge is to maintain a high level of effectiveness with the milder process, which is accomplished by using enzymes to further break down the hemicellulose after pretreatment. NREL has shown that proper mixtures of enzymes can enhance hemicellulose hydrolysis. In an experiment on pretreated corn stover, adding a hemicellulase enzyme to break down the hemicellulose increased the yield of xylose (a sugar resulting from hemicellulose hydrolysis) by 12% across a range of pretreatment conditions. Breaking down the hemicellulose also enhanced cellulose hydrolysis, resulting in a 6% higher glucose yield.

To get a broader look at pretreatment options, NREL participates in the Biomass Refining Consortium for Applied Fundamentals and Innovation (CAFI). Each CAFI participant is evaluating a different pretreatment approach using standardized experimental design and data reporting protocols. The CAFI projects allow participants to compare pretreatment and downstream process performance across a range of lignocellulosic feedstocks. NREL is using this knowledge to help identify the best pretreatment approaches for near- and long-term biorefining platforms. (For more information on CAFI, see "Coordinated Development of Leading Biomass Pretreatment Technologies," *Bioresource Technology*, December 2005.)

This pretreatment reactor hydrolyzes hemicellulose and solubilizes some lignin. The pressurized hot wash process separates these materials before they can reprecipitate. This system uses dilute sulfuric acid at increased temperature and pressure, but pressurized hot wash may work well with any pretreatment system.

ENZYMATIC HYDROLYSIS

Unlocking the Full Potential of Cellulosic Biomass

Plants have evolved over several hundred million years to be recalcitrant—resistant to attacks from the likes of bacteria, fungi, insects, and extreme weather. Breaking down plants is no easy task. For cellulosic ethanol production, the primary challenge is breaking down (hydrolyzing) cellulose into its component sugars.

NREL is exploring the causes of biomass recalcitrance and ways to overcome it using cellulases (enzymes that break down cellulose). The goals are to maximize the conversion of cellulose to sugar, accelerate the rate of conversion, and use fewer, cheaper enzymes. NREL's recent advances include employing state-of-the-art capabilities to characterize plant structure and developing superior enzymatic hydrolysis processes.

To make contact with cellulose, the enzymes must get past a complex maze of plant structures. NREL is mapping this labyrinth as a first step toward overcoming it. A unique array of microscopy tools and techniques in NREL's new Biomass Surface Characterization Laboratory enables researchers to image plant structures down to the molecular level. To probe even further—visualizing structures and processes at scales that cannot (yet) be observed—NREL and its partners are building a sophisticated molecular dynamics model of the cellulose-cellulase system. When complete, it will be the largest biological computer model ever developed.

Advanced computer modeling capabilities help NREL understand and improve enzymatic hydrolysis processes. This illustration of the CBH I enzyme is based on models developed by NREL.

 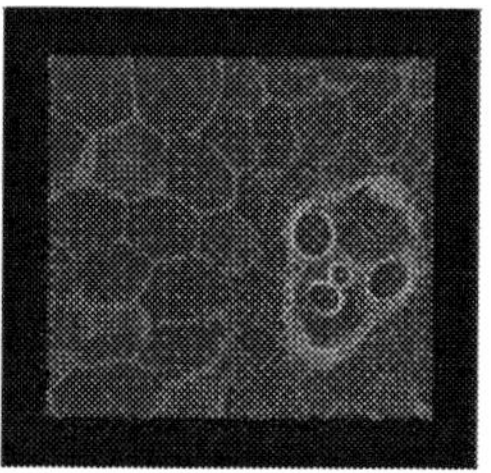 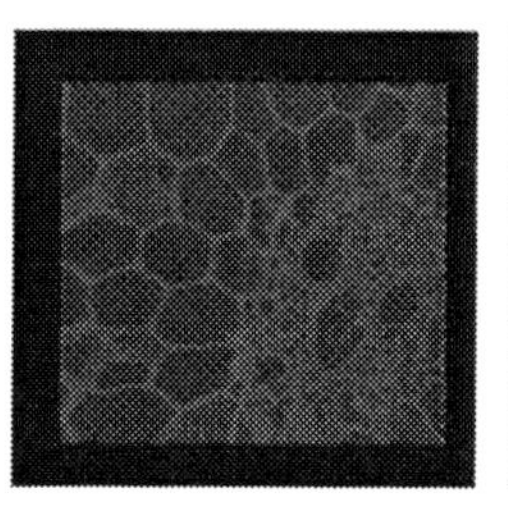 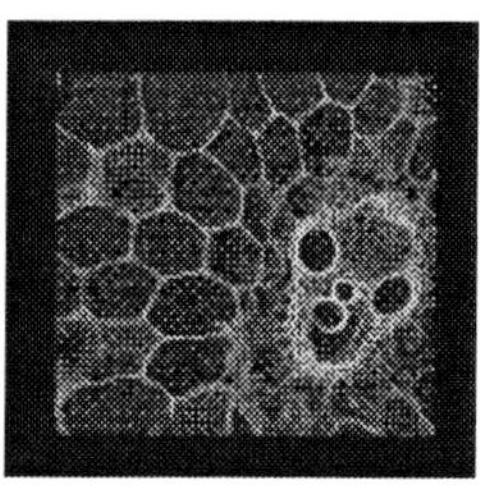

These ultra-sharp laser microscope images were created with the Biomass Surface Characterization Laboratory's scanning confocal microscope, which can be used to build 3-D representations of plant structures. (S.Porter, NREL)

Once cellulases make contact with cellulose, the real work begins. Cellulases act very slowly. That's why dead trees take years to decompose in the forest. To accelerate cellulose conversion, it is critical to start with the best enzymes nature has to offer. The most active known cellulases are in the cellobiohydrolase I (CBH I) family, derived from fungi. But not all CBH I enzymes are equal. NREL recently confirmed the existence of CBH I enzymes that are twice as active as those from industrial sources.

NREL and its partners Genencor International and Novozymes have developed a "cocktail" of cellulases to improve hydrolysis. In combination with NREL's process development improvements, this advance has reduced enzyme cost twentyfold. This work received an R&D 100 Award in 2004.

FERMENTATION

Creating "Super-Bugs" for Superior Ethanol Yield

During fermentation, microorganisms (primarily fungi and bacteria) convert the sugars in biomass to ethanol. Under ideal conditions, these "bugs" will work contentedly, consuming sugars and producing ethanol and other products. But conditions in a cellulosic ethanol biorefinery are anything but ideal.

The hot soup—called a hydrolyzate—generated after pretreatment and hydrolysis contains not only fermentable sugars, but also compounds (such as acetic acid) that are toxic to the bugs. Other things that are toxic in the fermentation process and the hydrolyzate are a high-solids concentration and a rising ethanol concentration. Because microorganisms found in nature do not function well in this hostile environment, NREL is creating "super-bugs" that thrive in it.

Yeasts are currently the fermentation organisms of choice for the corn ethanol industry. They are reasonably tolerant of ethanol, acid, and moderately high temperatures. However, existing yeast strains cannot withstand highly toxic hydrolyzates or ferment 5-carbon sugars and minor 6-carbon sugars efficiently. NREL, along with the National Corn Growers Association (NCGA) and Corn Refiners Association (CRA), developed yeast capable of fermenting a particular 5-carbon sugar, arabinose, which constitutes up to 20% of the fermentable sugars in corn fiber. Three genes from a bacterium were inserted into the yeast Saccharomyces cerevisiae. This work resulted in the first ever demonstration, in 2000, of arabinose fermentation by yeast. Next, NREL plans to test the strain under real biorefining conditions—in the hydrolyzate.

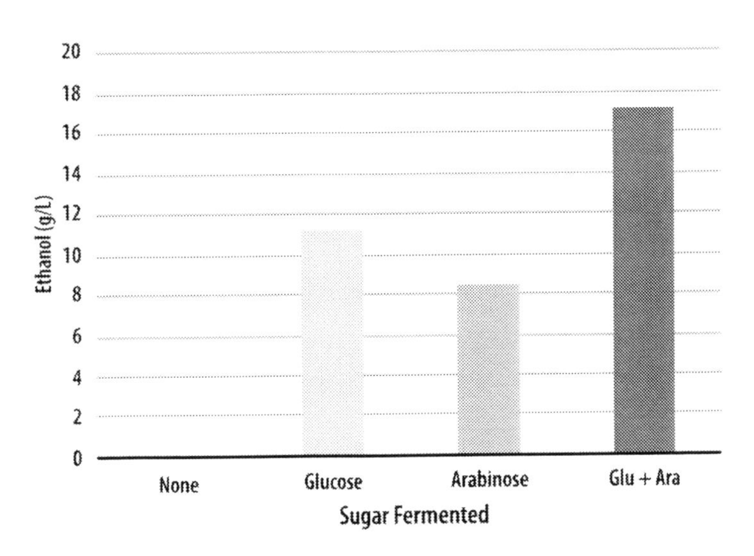

The yeast strain developed by NREL, NCGA, and CRA is the world's first yeast to ferment arabinose. These results show an ethanol yield of 83% from arabinose in a defined medium (not a hydrolyzate). From left to right, initial sugar concentrations were 0 g/L, 20 g/L glucose, 20 g/L arabinose, and 20 g/L glucose + 20 g/L arabinose. Expected ethanol from 20 g/L of sugar is 10.2 g/L at 100% yield. (A. Singh, NREL, patent pending)

Fermentation processes are tested in NREL's biochemical process development unit (PDU). This 9,000-L fermenter is large enough to produce sufficient lignin for processing in the downstream thermochemical PDU. Warren Gretz, NREL PIX 00945

NREL also pioneered the use of a yeast alternative, the bacterium Zymomonas mobilis (Zymo). Zymo gives a high ethanol yield and tolerates high ethanol concentrations. Using genetic and metabolic engineering, NREL developed acetic acid-tolerant Zymo strains that can ferment arabinose and the most important 5-carbon sugar, xylose. This strain resulted in several patents and an R&D 100 Award. NREL also pioneered a technique to make the Zymo strain stable (the bacteria's offspring have the same genes as the parents) by inserting key genes into the genome. NREL's Zymo work has included successful collaborations with the NCGA and CRA, the chemical company Arkenol (now BlueFire Ethanol), and DuPont.

PROCESS INTEGRATION

Tying Together the Integrated Biorefinery

To produce low-cost ethanol, biorefineries will need to link the refining steps into an integrated process. However, optimizing conditions in one step of the process can influence performance in other steps. The challenge is to find the right combination of trade-offs that optimize the integrated process. Studying integrated biorefinery operations requires an advanced process development unit and state-of-the-art chemical analysis capabilities. NREL's research on high-solids operation is a good example.

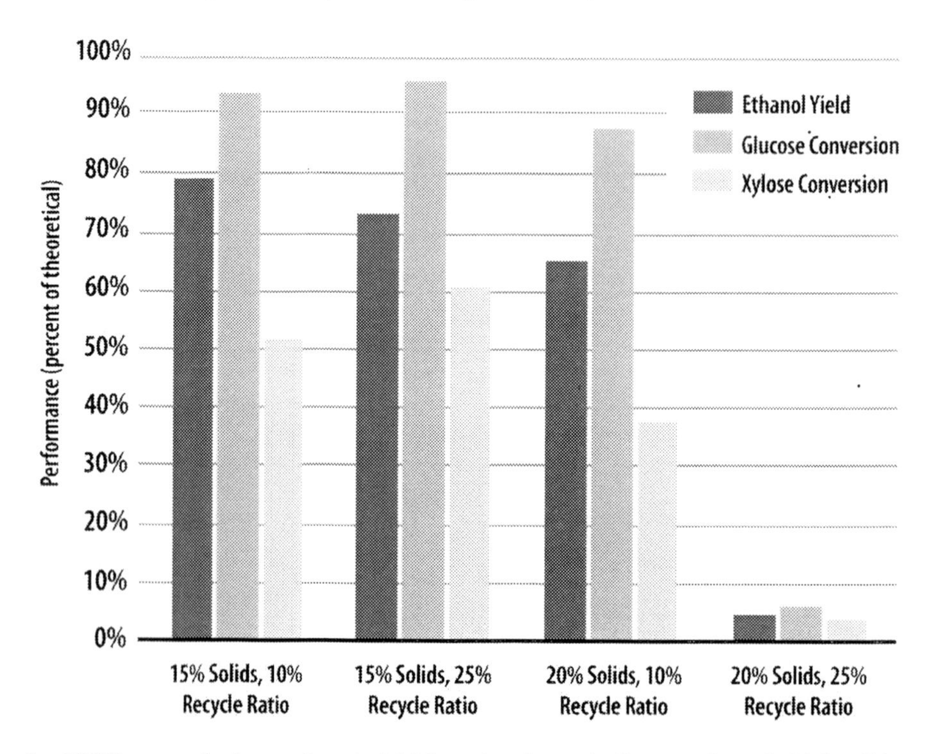

Pioneering NREL research shows ethanol yield dropping dramatically at moderately high solids concentration and high recycle ratio (the ratio of recycled water to fresh water). (D. Schell, NREL)

NREL has shown that high-solids operation—using a high ratio of biomass to water in the biorefining process—is one key to cutting ethanol costs. The less water introduced in the pretreatment step, the higher the potential sugar concentration and the less equipment and energy the process requires. The result is lower-cost ethanol. In a perfect world that would be enough, but high solids concentrations can create problems elsewhere in the process.

In a first-of-its-kind study, NREL demonstrated that a moderately high solids concentration combined with recycled process water severely inhibits fermentation and, consequently, lowers ethanol yield. This is important because commercial biorefineries will need to recycle water, taking it from the back end of the refining process and combining it with fresh water at the front end of the process. NREL's study identified the ability to achieve high solids concentration and high levels of process water recycle as an issue that must be considered in both pretreatment and fermentative microorganism development.

Managing the properties of high-solids mixtures is another process integration challenge. Like adding flour to water in a recipe, adding biomass to water makes the mixture thicker and more viscous. This has important implications for the efficient flow and conversion of biomass through the integrated process. NREL is developing unique capabilities in biomass rheology—the science of the deformation and flow of materials—to determine the best ways to manage high-solids biomass mixtures. This research is critical to providing process engineers with rheological information needed to design a commercial biorefinery.

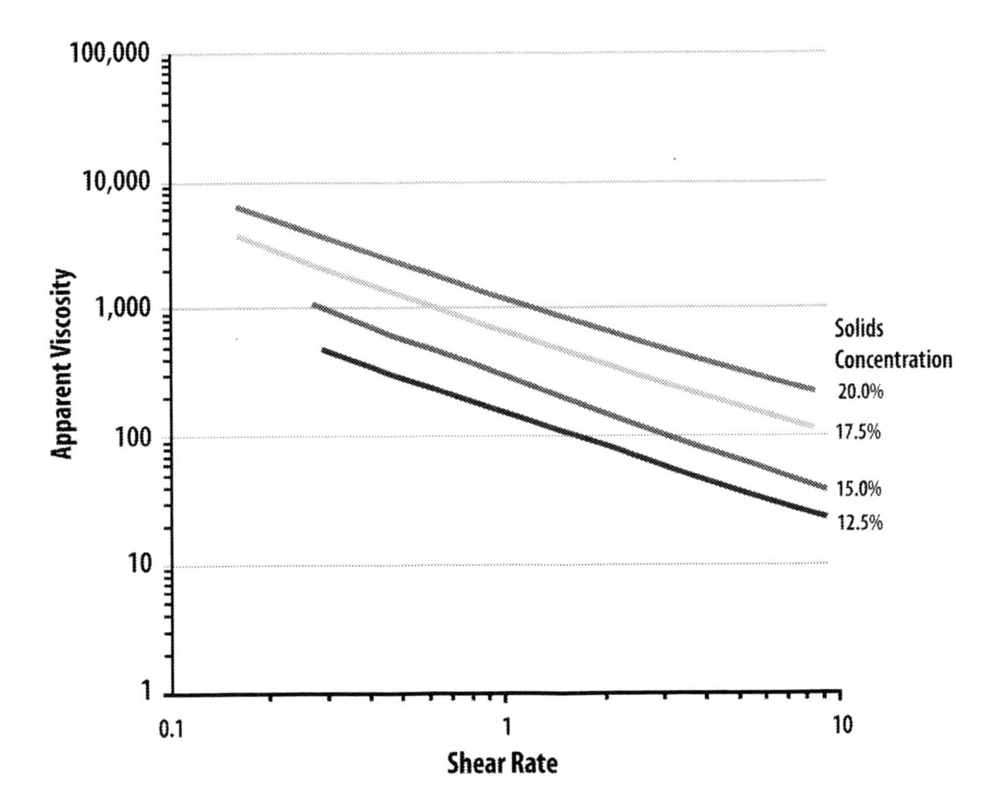

NREL is quantifying the rheological properties of various biomass materials. These results show how the viscosity (resistance to flow) of pretreated corn stover increases with higher solids concentration and decreases with faster shear rate (speed of mixing). (J. McMillan, NREL)

THERMOCHEMICAL CONVERSION

Honing a Powerful Path to Economic Ethanol

The advances described up to this point relate to biochemical conversion of carbohydrates to ethanol. However, ethanol can also be produced thermochemically from any form of biomass. In this approach, heat and chemicals are used to break biomass into syngas (CO and H2) and reassemble it into products such as ethanol. This method is particularly important because up to one third of cellulosic biomass—the lignin-rich parts—cannot be easily converted biochemically.

Forest products and mill residues typically have high lignin contents, making them unattractive feedstocks for biochemical conversion yet suitable for thermochemical conversion. In an integrated biorefinery, lignin-rich residues from the biochemical process could also be converted thermochemically.

A thermochemical biomass conversion process is complex, and uses components, configurations, and operating conditions that are more typical of petroleum refining. And, just like researchers in the petroleum industry, NREL uses a combination of experimental research together with process economic models to explore a large number of possible process configurations. A much simplified schematic of NREL's preferred configuration is shown below. This configuration employs an indirect gasifier, tar reforming, and a mixed alcohol synthesis step designed to maximize ethanol yield.

In thermochemical conversion, biomass is converted into syngas, and syngas is converted into an ethanol-rich mixture. However, syngas created from biomass is not "clean"—it contains contaminants such as tar and sulfur that interfere with the conversion of the syngas into products. These contaminants must be removed. NREL has developed tar-reforming catalysts and catalytic reforming processes that have demonstrated high levels of tar conversion—converting up to 97% of the tar into more syngas. This not only cleans the syngas, it also creates more of it, improving process economics and ultimately cutting the cost of the resulting ethanol. NREL has also made progress regenerating the tar-reforming catalyst after it has been partially deactivated by sulfur poisoning.

NREL is evaluating many different process options and their associated costs to help identify key barriers to low-cost ethanol production. For example, process models indicate that reducing tars and hydrocarbons from syngas can decrease the production cost of ethanol by 33%. NREL models also highlight the need for extensive heat integration and quantify performance targets needed to achieve DOE's ethanol cost goals thermochemically.

BIOMASS ANALYSIS

Enabling Total Control of Ethanol Production

You can't control what you can't measure. Biorefinery operators will need to know the precise composition of the biomass going through their processes so they can tightly control the cost and quality of the products coming out. The faster and more reliable the measurements are, the better and cheaper the final products will be.

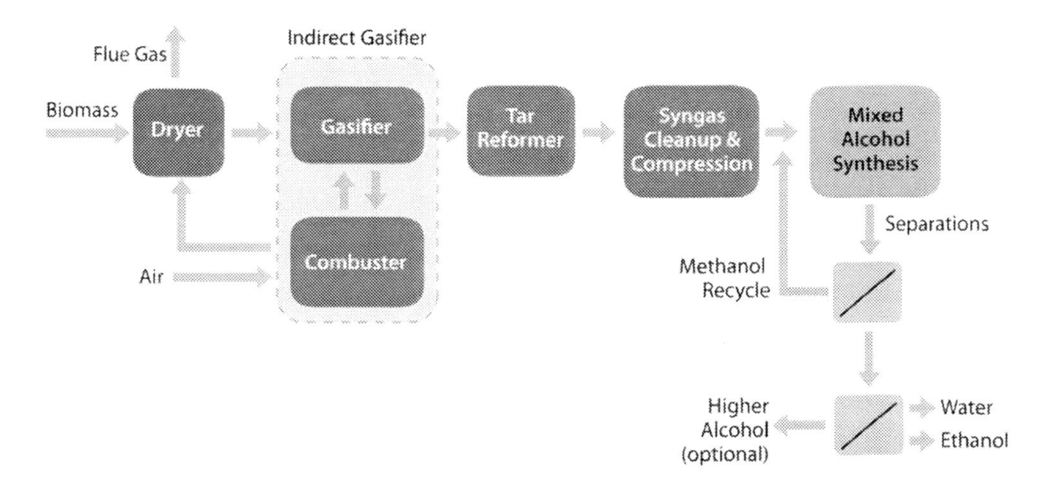

This simplified schematic of NREL's thermochemical conversion model shows the many steps that can be manipulated to optimize efficiency and cost.

NREL's capabilities are constantly evolving to meet industry's need for accurate and rapid biomass analysis. Researchers recently developed a way to rapidly analyze biomass composition using near-infrared (NIR) spectroscopy. In this technique, light reflected off a biomass sample is analyzed to determine the sample's composition. Compared with traditional wet chemistry analysis, this is a huge leap forward. Analyzing a sample with this new approach takes minutes instead of weeks and costs tens of dollars instead of thousands of dollars—without sacrificing precision or accuracy. The rugged NIR instruments can even be adapted to a working biorefinery to measure, for example, the chemical composition of corn stover as it enters and exits pretreatment.

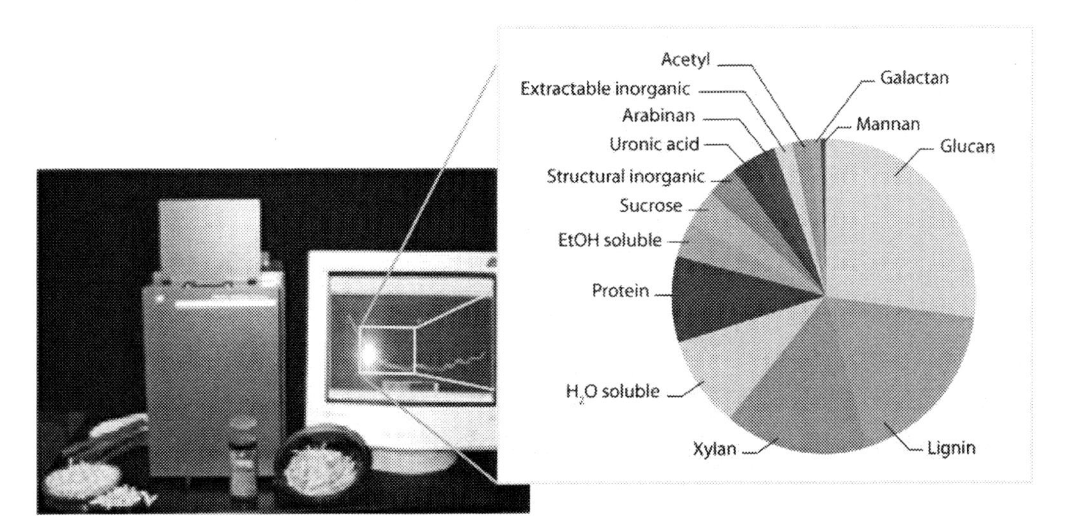

NREL's NIR spectroscopy technique rapidly analyzes biomass composition and can be adapted to on-line use.

What if you could measure an adult plant's composition while it is still a sprout? You could grow plants that yielded the most ethanol, maximize cellulose content, and minimize lignin. Or what if, by knowing how genes affect plant composition, you could create the ideal ethanol feedstock? NREL's analytical capabilities are making these scenarios a reality. Using techniques such as molecular beam mass spectrometry and nuclear magnetic resonance spectroscopy, NREL is measuring cell wall chemistry more quickly and accurately than anyone else in the world. These measurements are used to predict adult plant composition. In 2006, NREL analyzed the cell wall chemistry of more than 10,000 samples for industry and university partners. These thermoanalytical capabilities can help accelerate crop breeding and genetic engineering.

CONTACTS

For more information about working with NREL, please contact:
John Ashworth, Technical Lead for Partnerships and Contracts, (303) 384-6858

Please see the following Web sites for more information on biomass and related research at NREL:

*NREL's R&D "Working With Us" Web page:*www.nrel.gov/biomass/workingwithus.html
NREL's Biomass Research Web site: www.nrel.gov/biomass
NREL's Chemical & Biosciences Center Web site: www.nrel.gov/basic_sciences/
DOE Biomass Program Web site: www.eere.energy.gov/biomass/

The National Bioenergy Center is headquartered at NREL:
National Renewable Energy Laboratory 1617 Cole Blvd.
Golden, CO 80401 303.275.3000 • www.nrel.gov
Operated for the U.S. Department of Energy Office of Energy Efficiency and Renewable Energy by Midwest Research Institute • Batelle

In: Ethanol Biofuel Production
Editor: Bratt P. Haas
ISBN: 978-1-60876-086-2

Chapter 3

THERMOCHEMICAL ETHANOL VIA INDIRECT GASIFICATION AND MIXED ALCOHOL SYNTHESIS OF LIGNOCELLULOSIC BIOMASS

S. Phillips, A. Aden, J. Jechura and D. Dayton

1. EXECUTIVE SUMMARY

This work addresses a policy initiative by the Federal Administration to apply United States Department of Energy (DOE) research to broadening the country's domestic production of economic, flexible, and secure sources of energy fuels. President Bush stated in his 2006 State of the Union Address: "America is addicted to oil." To reduce the Nation's future demand for oil, the President has proposed the Advanced Energy Initiative which outlines significant new investments and policies to change the way we fuel our vehicles and change the way we power our homes and businesses. The specific goal for biomass in the Advanced Energy Initiative is to foster the breakthrough technologies needed to make cellulosic ethanol cost-competitive with corn-based ethanol by 2012.

In previous biomass conversion design reports by the National Renewable Energy Laboratory (NREL), a benchmark for achieving production of ethanol from cellulosic feedstocks that would be "cost competitive with corn-ethanol" has been quantified as $1.07 per gallon ethanol minimum plant gate price.

This process design and technoeconomic evaluation addresses the conversion of biomass to ethanol via thermochemical pathways that are expected to be demonstrated at the pilot-unit level by 2012. This assessment is unique in its attempt to match up:

- Currently established and published technology.
- Technology currently under development or shortly to be under development from DOE Office of Biomass Program funding.
- Biomass resource availability in the 2012 time frame consistent with the Billion Ton Vision study.

Indirect steam gasification was chosen as the technology around which this process was developed based upon previous technoeconomic studies for the production of methanol and hydrogen from biomass. The operations for ethanol production are very similar to those for methanol production (although the specific process configuration will be different). The general process areas include: feed preparation, gasification, gas cleanup and conditioning, and alcohol synthesis & purification.

The cost of ethanol as determined in this assessment was derived using technology that has been developed and demonstrated or is currently being developed as part of the OBP research program. Combined, all process, market, and financial targets in the design represent what must be achieved to obtain the reported $1.01 per gallon, showing that ethanol from a thermochemical conversion process has the possibility of being produced in a manner that is "cost competitive with corn-ethanol" by 2012. This analysis has demonstrated that forest resources can be converted to ethanol in a cost competitive manner. This allows for greater flexibility in converting biomass resources to make stated volume targets by 2030.

2. INTRODUCTION

This work addresses a policy initiative by the Federal Administration to apply United States Department of Energy (DOE) research to broadening the country's domestic production of economic, flexible, and secure sources of energy fuels. President Bush stated in his 2006 State of the Union Address: "America is addicted to oil." [1] To reduce the Nation's future demand for oil, the President has proposed the Advanced Energy Initiative [2] which outlines significant new investments and policies to change the way we fuel our vehicles and change the way we power our homes and businesses. The specific goal for biomass in the Advanced Energy Initiative is to foster the breakthrough technologies needed to make cellulosic ethanol cost- competitive with corn-based ethanol by 2012.

In previous biomass conversion design reports by the National Renewable Energy Laboratory (NREL), a benchmark for achieving production of ethanol from cellulosic feedstocks that would be "cost competitive with corn-ethanol" has been quantified as $1.07 per gallon ethanol minimum plant gate price [3] (where none of these values have been adjusted to a common cost year). The value can be put in context with the historic ethanol price data as shown in Figure 1 [4]. The $1.07 per gallon value represents the low side of the historical fuel ethanol prices. Given this historical price data, it is viewed that cellulosic ethanol would be commercially viable if it was able to meet a minimum return on investment selling at this price.

This is a cost target for this technology; it does not reflect NREL's assessment of where the technology is today. Throughout this report, two types of data will be shown: results which have been achieved presently in a laboratory or pilot plant, and results that are being targeted for technology improvement several years into the future. Only those targeted for the 2012 timeframe are included in this economic evaluation. Other economic analyses that attempt to reflect the current "state of technology" are not presented here.

Conceptual process designs and associated design reports have previously been done by NREL for converting cellulosic biomass feedstock to ethanol via Biochemical pathways. Two types of biomass considered have been yellow poplar [5] and corn stover. [3] These design reports have been useful to NREL and DOE program management for two main reasons. First of all, they *enable comparison of research and development projects.* A conceptual process design helps to direct research by establishing a benchmark to which other process configurations can be compared. The anticipated results of proposed research can be translated into design changes; the economic impact of these changes can then be determined and this new design can be compared to the benchmark case. Following this procedure for several proposed research projects allows DOE to make competitive funding decisions based on which projects have the greatest potential to lower the cost of ethanol production. Complete process design and economics are required for such comparisons because changes in performance in one research area may have significant impacts in other process areas not part of that research program (e.g., impacts in product recovery or waste treatment). The impacts on the other areas may have significant and unexpected impacts on the overall economics.

Secondly, they enable *comparison of ethanol production to other fuels.* A cost of production has also been useful to study the potential ethanol market penetration from technologies to convert lignocellulosic biomass to ethanol. The cost estimates developed must be consistent with applicable engineering, construction, and operating practices for facilities of this type. The complete process (including not only industry-standard process components but also the newly researched areas) must be designed and their costs determined.

Following the methodology of the biochemical design reports, this process design and technoeconomic evaluation addresses the conversion of biomass to ethanol via thermochemical (TC) pathways that are expected to be demonstrated at the pilot-unit level by 2012. This assessment is unique in its attempt to match up:

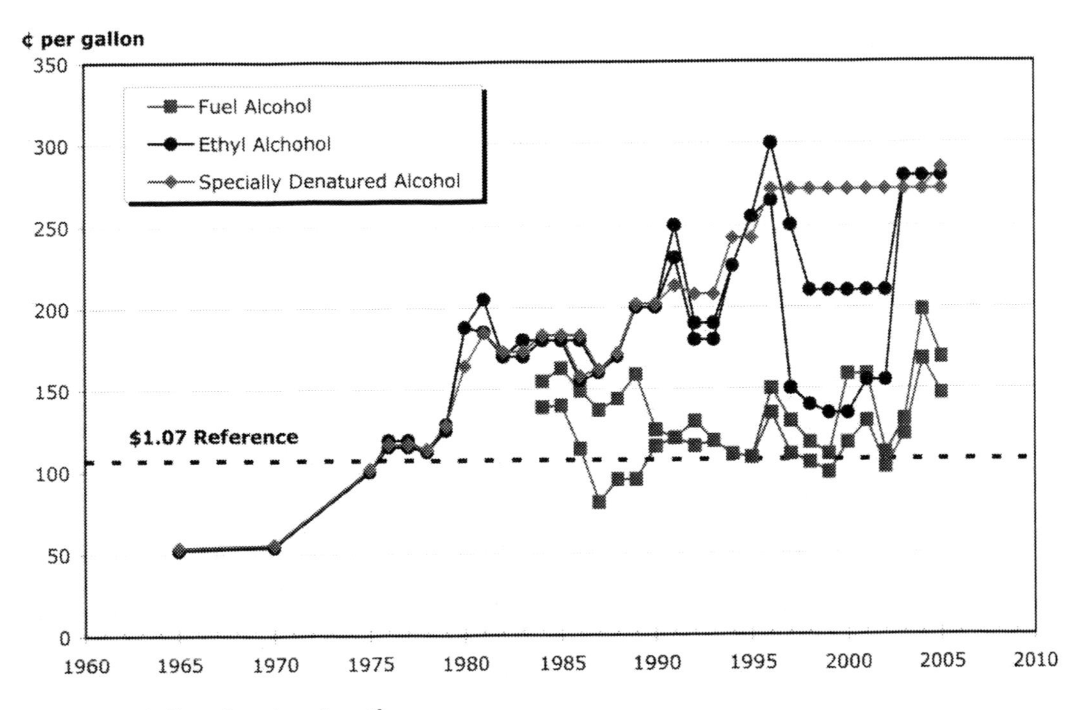

Figure 1. U.S. list prices for ethanol[a]

- Currently established and published technology.
- Technology currently under development or shortly to be under development from DOE Office of Biomass Program (OBP) funding. (See Appendix B for these research targets and values.)
- Biomass resource availability in the 2012 time frame consistent with the Billion Ton Vision study [6].

This process design and associated report provides a benchmark for the Thermochemical Platform just as the Aden et al. report [3] has been used as a benchmark for the Biochemical Platform since 2002. It is also complementary to gasification-based conversion assessments done by NREL and others. This assessment directly builds upon an initial analysis for the TC production of ethanol and other alcohol co-products [7, 8], which, in turn, was based upon a detailed design and economic analysis for the production of hydrogen from biomass.[9] This design report is also complementary to other studies being funded by the DOE OBP, including the RBAEF (Role of Biomass in America's Energy Future) study [10]. However, the RBAEF study differs in many ways from this study. For example, RBAEF is designed for a further time horizon than 2012. It is based on a different feedstock, switchgrass, and it considers a variety of thermochemical product options, including ethanol, power and Fischer-Tropsch liquids [11]. Other notable gasification studies have been completed by Larsen at Princeton University, including a study examining the bioproduct potential of Kraft mill black liquor based upon gasification [12].

Indirect steam gasification was chosen as the technology around which this process was developed based upon previous technoeconomic studies for the production of methanol and hydrogen from biomass [13]. The sub-process operations for ethanol production are very similar to those for methanol production (although the specific process configuration will be different). The general process areas include: feed preparation, gasification, gas cleanup and conditioning, and alcohol synthesis & purification.

Gasification involves the devolatilization and conversion of biomass in an atmosphere of steam and/or oxygen to produce a medium-calorific value gas. There are two general classes of gasifiers. *Partial oxidation (PDX)* gasifiers (directly-heated gasifiers) use the exothermic reaction between oxygen and organics to provide the heat necessary to devolatilize biomass and to convert residual carbon-rich chars. In POX gasifiers, the heat to drive the process is generated internally within the gasifier. A disadvantage of POX gasifiers is that oxygen production is expensive and typically requires large plant sizes to improve economics [14].

The second general class, *steam gasifiers* (indirectly-heated gasifiers), accomplish biomass heating and gasification through heat transfer from a hot solid or through a heat transfer surface. Either byproduct char and/or a portion of the product gas can be combusted with air (external to the gasifier itself) to provide the energy required for gasification. Steam gasifiers have the advantage of not requiring oxygen; but since most operate at low pressure they require product gas compression for downstream purification and synthesis unit operations. The erosion of refractory due to circulating hot solids in an indirect gasifier can also present some potential operational difficulties.

A number of POX and steam gasifiers are under development and have the potential to produce a synthesis gas suitable for liquid fuel synthesis. These gasifiers have been operated in the 4 to 350 ton per day scale. The decision as to which type of gasifier (POX or steam) will be the most economic depends upon the entire process, not just the cost for the gasifier itself. One indicator for comparing processes is "capital intensity," the capital cost required on a per unit product basis. Figure 2 shows the capital intensity of methanol processes [15, 16, 17, 18, 19, 20] based on indirect steam gasification and direct POX gasification. This figure shows that steam gasification capital intensity is comparable or lower than POX gasification. The estimates indicate that both steam gasification and POX gasification processes should be evaluated, but if the processes need to be evaluated sequentially, choosing steam gasification for the first evaluation is reasonable.

Another philosophy applied to the process development was the idea to make the process energy self-sufficient. It was recognized that the heat and power requirements of the process could not be met just with char combustion and would require additional fuel. Several options were considered. Additional biomass could be added as fuel directly to the heat and power system, however, this would increase the process beyond 2,000 tonne/day. Fossil fuels (coal or natural gas) could also be added directly to provide the additional fuel. Alternately syngas could be diverted from liquid fuel production to heat and power production. This option makes the design more energy self-sufficient, but also lowers the overall process yield of alcohols.

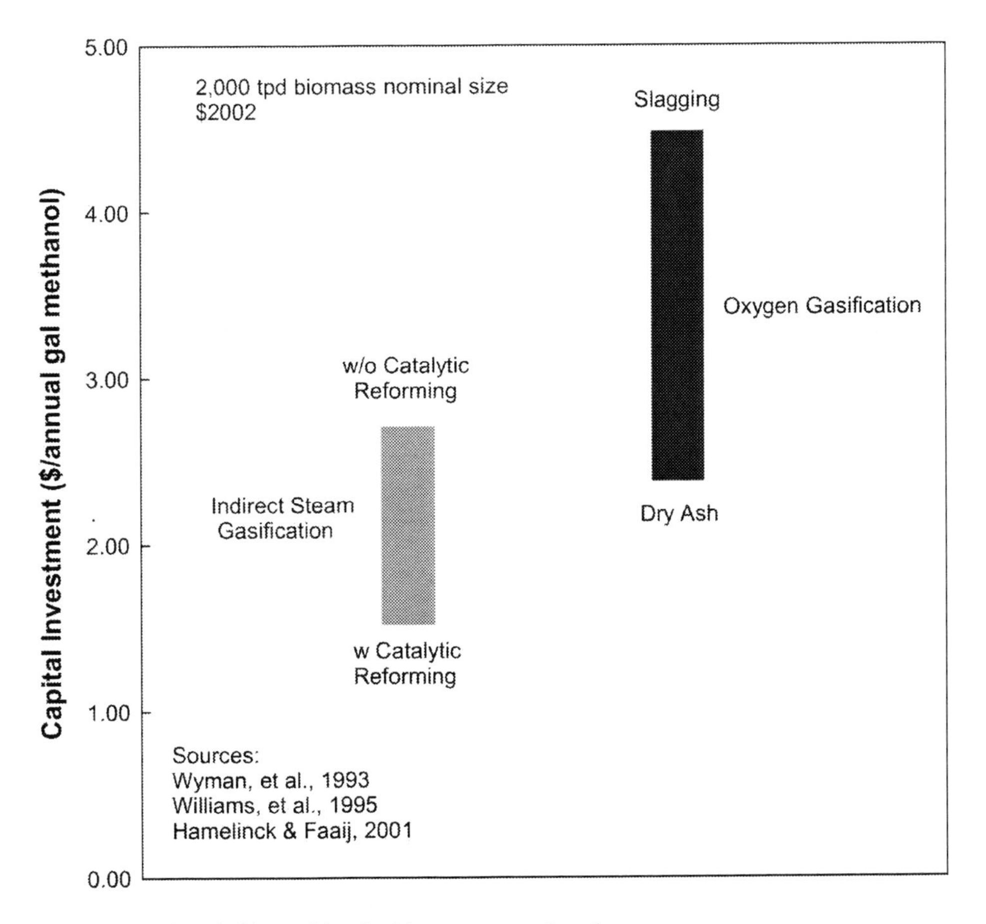

Figure 2. Estimated capital intensities for biomass-to-methanol processes

It was decided that (1) no additional fuel would be used for heat and power and (2) only enough syngas would be diverted so that the internal heat and power requirements would be exactly met. Thus, there would neither be electricity sales to the grid nor electricity purchases. The only exception to this would be if other operating specifications were such that syngas could no longer be backed out of the heat and power system but there is still excess electricity (that could then be sold to the grid for a co-product credit). This resulted in 28% of the unconditioned syngas being diverted to power the process. Model calculations show that if none of the syngas was diverted in this manner, and all of it was used for mixed alcohols production, the ethanol and higher alcohols yields would increase by 38%. Thus, the baseline ethanol yield of 80.1 gal/dry ton could rise as high as 110.9 gal/ton, with total production of all alcohols as high as 130.3 gal/dry ton. However, the minimum ethanol plant gate price increases in this scenario because of the cost of the natural gas required to meet the energy demands of the process.

2.1. Analysis Approach

The general approach used in the development of the process design, process model, and economic analysis is depicted in Figure 3. The first step was to assemble a general process flow schematic or more detailed process flow diagrams (PFDs). (See Appendix H for the associated PFDs for this design). From this, detailed mass and energy balance calculations were performed around the process. For this design, Aspen Plus software was used. Data from this model was then used to properly size all process equipment and fully develop an estimate of capital and operating costs. These costs could have potentially been used in several types of economic analysis. For this design however, a discounted cash flow rate of return (DCFROR) analysis was used to determine the ethanol minimum plant gate price necessary to meet an nth plant hurdle rate (IRR) of 10%.

This TC conversion process was developed based upon NREL experience performing conceptual designs for biomass conversion to ethanol via biochemical means [3], biopower applications, and biomass gasification for hydrogen production.[9] Specific information for potential subprocesses were obtained as a result of a subcontract with Nexant Inc. [21, 22, 23, 24]

Aspen Plus version 2004.1 was used to determine the mass and energy balances for the process. The operations were separated into seven major HIERARCHY areas:

- Feed Handling and Drying (Area 100)
- Gasification (Area 200)
- Cleanup and Conditioning (Area 300)
- Alcohol Synthesis (Area 400)
- Alcohol Separation (Area 500)
- Steam Cycle (Area 600)
- Cooling Water (Area 700)

Overall, the Aspen simulation consists of about 300 operation blocks (such as reactors, flash separators, etc.), 780 streams (mass, heat, and work), and 65 control blocks (design specs and calculator blocks). Many of the gaseous and liquid components were described as distinct molecular species using Aspen's own component properties database. The raw biomass feedstock, ash, and char components were modeled as non-conventional components. There was more detail and rigor in some blocks (e.g., distillation columns) than others (e.g., conversion extent in the alcohol synthesis reactor). Because this design processes three different phases of matter (solid phase, gas phase, and liquid phase), no single thermodynamics package was sufficient. Instead, four thermodynamics packages were used within the Aspen simulation to give more appropriate behavior. The "RKS-BM" option was used throughout much of the process for high temperature, high pressure phase behavior. The non-random two-liquid "NRTL" option with ideal gas properties was used for alcohol separation calculations. The 1987 Steam Table properties were used for the steam cycle calculations. Finally, the ELECNRTL package was used to model the electrolyte species potentially present within the quench water system.

The process economics are based on the assumption that this is the "nth" plant, meaning that several plants using this same technology will have already been built and are operating. This means that additional costs for risk financing, longer start-ups, and other costs associated with first-of-a-kind plants are not included.

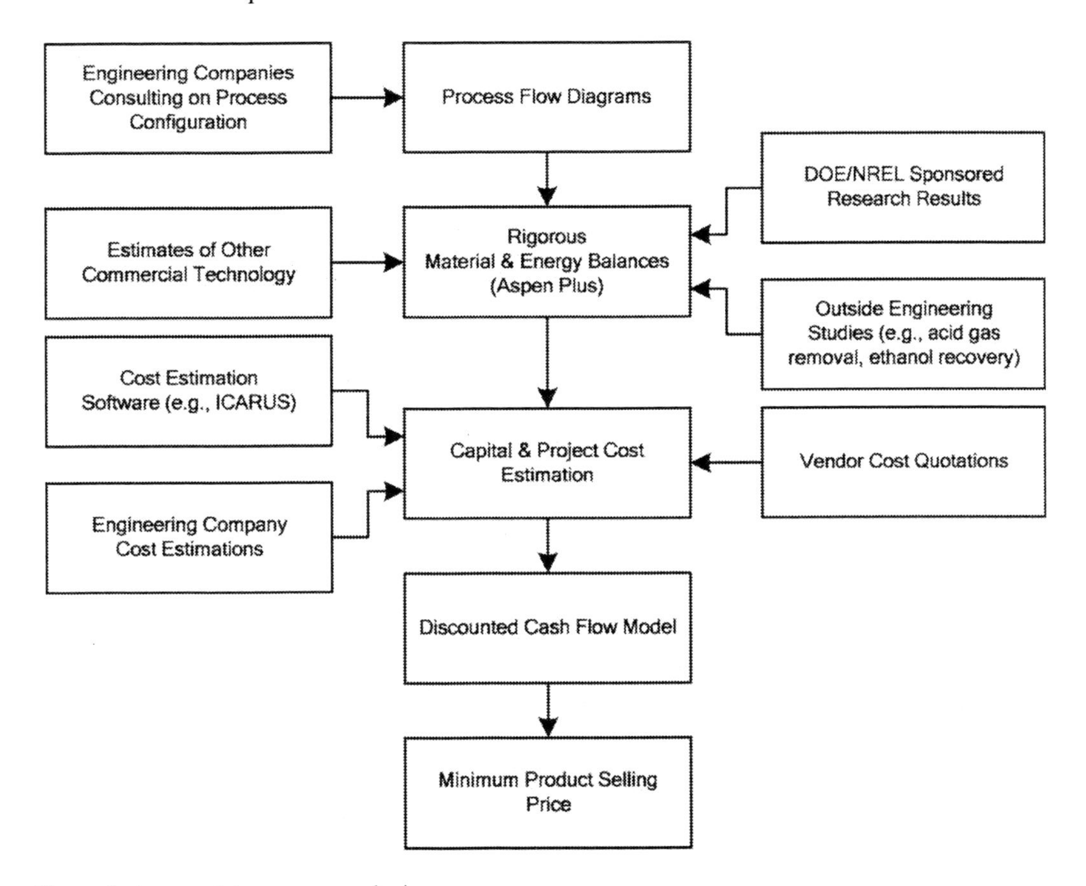

Figure 3. Approach to process analysis

The capital costs were developed from a variety of sources. For some sub-processes that are well known technology and can be purchased as modular packages (i.e. amine treatment, acid gas removal), an overall cost for the package unit was used. Many of the common equipment items (tanks, pumps, simple heat exchangers) were costed using the Aspen Icarus Questimate costing software. Other more specific unit operations (gasifier, molecular sieve, etc) used cost estimates from other studies and/or from vendor quotes. As documented in the hydrogen design report [9], the installed capital costs were developed using general plant-wide factors. The installation costs incorporated cost contributions for not only the actual installation of the purchased equipment but also instrumentation and controls, piping, electrical systems, buildings, yard improvements, etc. These are also described in more detail in Section 3.

The purchased component equipment costs reflect the base case for equipment size and cost year. The sizes needed in the process may actually be different than what was specifically designed. Instead of re-costing in detail, an exponential scaling expression was used to adjust the bare equipment costs:

$$\text{New Cost} = (\text{Base Cost})\left(\frac{\text{New Size}}{\text{Base Size}}\right)^n$$

where n is a characteristic scaling exponent (typically in the range of 0.6 to 0.7). The sizing parameters are based upon some characteristic of the equipment related to production capacity, such as inlet flow or heat duty in a heat exchanger (appropriate if the log-mean temperature difference is known not to change greatly). Generally these related characteristics are easier to calculate and give nearly the same result as resizing the equipment for each scenario. The scaling exponent n can be inferred from vendor quotes (if multiple quotes are given for different sizes), multiple estimates from Questimate at different sizes, or a standard reference (such as Garrett, [25] Peters and Timmerhaus, [26] or Perry et al. [27]).

Since a variety of sources were used, the bare equipment costs were derived based upon different cost years. Therefore, all capital costs were adjusted with the *Chemical Engineering* (CE) magazine's Plant Cost Index [28] to a common basis year of 2005:

$$\text{New Cost} = (\text{Base Cost})\left(\frac{\text{Cost Index in New Year}}{\text{Cost Index in Base Year}}\right).$$

The CE indices used in this study are listed in Table 1 and depicted in Figure 4. Notice that the indices were very nearly the same for 2000 to 2002 (essentially zero inflation) but take a very sharp increase after 2003 (primarily due a run-up in worldwide steel prices).

Once the scaled, installed equipment costs were determined, we applied overhead and contingency factors to determine a total plant investment cost. That cost, along with the plant operating expenses (generally developed from the ASPEN model's mass and energy balance results) was used in a discounted cash flow analysis to determine the ethanol plant gate price, using a specific discount rate. For the analysis done here, the ethanol minimum plant gate price is the primary value used to compare alternate designs.

Table 1. *Chemical Engineering* Magazine's Plant Cost Indices

Year	Index	Year	Index
1990	357.6	1998	389.5
1991	361.3	1999	390.6
1992	358.2	2000	394.1
1993	359.2	2001	394.3
1994	368.1	2002	395.6
1995	381.1	2003	402.0
1996	381.7	2004	444.2
1997	386.5	2005	468.2

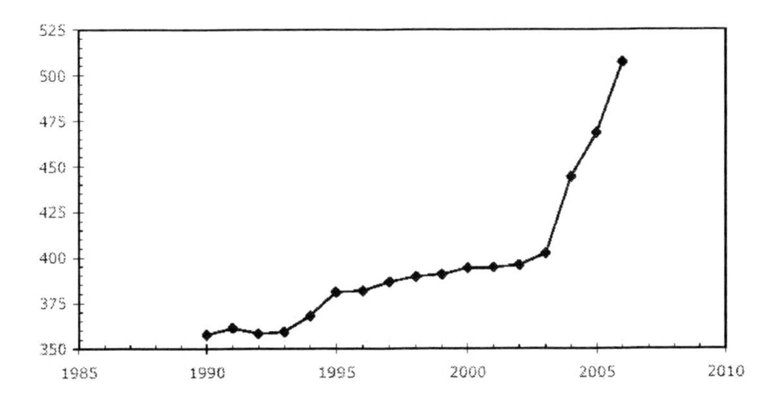

Figure 4. *Chemical Engineering* Magazine's plant cost indices

2.2. Process Design Overview

A simple block flow diagram of the current design is depicted in Figure 5. The detailed process flow diagrams (PFDs) are in Appendix H. The process has the following steps:

- *Feed Handling & Preparation.* The biomass feedstock is dried from the as-received moisture to that required for proper feeding into the gasifier using flue gases from the char combustor and tar reformer catalyst regenerator.

- *Gasification.* Indirect gasification is considered in this assessment. Heat for the endothermic gasification reactions is supplied by circulating hot synthetic olivine*[a] "sand" between the gasifier and the char combustor. Conveyors and hoppers are used to feed the biomass to the low-pressure indirectly-heated entrained flow gasifier. Steam is injected into the gasifier to aid in stabilizing the entrained flow of biomass and sand through the gasifier. The biomass chemically converts to a mixture of syngas components (CO, H_2, CO_2, CH_4, etc.), tars, and a solid "char" that is mainly the fixed carbon residual from the biomass plus carbon (coke) deposited on the sand. Cyclones at the exit of the gasifier separate the char and sand from the syngas. These solids flow by gravity from the cyclones into the char combustor. Air is introduced to the bottom of the reactor and serves as a carrier gas for the fluidized bed plus the oxidant for burning the char and coke. The heat of combustion heats the sand to over 1800°F. The hot sand and residual ash from the char is carried out of the combustor by the combustion gases and separated from the hot gases using another pair of cyclones. The first cyclone is designed to capture mostly sand while the smaller ash particles remain entrained in the gas exiting the cyclone. The second cyclone is designed to capture the ash and any sand passing through the first cyclone. The hot sand captured by the first cyclone flows by gravity back into the gasifier to provide the heat for the gasification reaction. Ash and sand particles captured in the second cyclone are cooled, moistened to minimize dust and sent to a land fill for disposal.

- *Gas Cleanup & Conditioning.* This consists of multiple operations: reforming of tars and other hydrocarbons to CO and H_2; syngas cooling/quench; and acid gas (CO_2 and H_2S) removal with subsequent reduction of H2S to sulfur. Tar reforming is envisioned to occur in an isothermal fluidized bed reactor; de-activated reforming catalyst is separated from the effluent syngas and regenerated on-line. The hot syngas is cooled through heat exchange with the steam cycle and additional cooling via water scrubbing. The scrubber also removes impurities such as particulates and ammonia along with any residual tars. The excess scrubber water is sent off-site to a waste-water treatment facility. The cooled syngas enters an amine unit to remove the CO_2 and H_2S. The H2S is reduced to elemental sulfur and stockpiled for disposal. The CO_2 is vented to the atmosphere in this design.

- *Alcohol Synthesis.* The cleaned and conditioned syngas is converted to alcohols in a fixed bed reactor. The mixture of alcohol and unconverted syngas is cooled through heat exchange with the steam cycle and other process streams. The liquid alcohols are separated by condensing them away from the unconverted syngas. Though the unconverted syngas has the potential to be recycled back to the entrance of the alcohol synthesis reactor, this recycle is not done in this process design because CO_2 concentrations in the recycle loop would increase beyond acceptable limits of the catalyst. Added cost would be incurred if this CO_2 were separated. Instead the unconverted syngas is recycled to the Gas Cleanup & Conditioning section, mostly as feed to the tar reformer.

- *Alcohol Separation.* The alcohol stream from the Alcohol Synthesis section is depressurized in preparation of dehydration and separation. Another rough separation is performed in a flash separator; the evolved syngas is recycled to the Gas Cleanup & Conditioning section, mostly as feed to the tar reformer. The depressurized alcohol stream is dehydrated using vapor-phase molecular sieves. The dehydrated alcohol stream is introduced to the main alcohol separation column that splits methanol and ethanol from the higher molecular weight alcohols. The overheads are topped in a second column to remove the methanol to ASTM sales specifications. The methanol leaving in the overheads is used to flush the adsorbed water from the molecular sieves. This methanol/water mixture is recycled back to the entrance of the alcohol synthesis reactor in order to increase the yield of ethanol and higher alcohols.

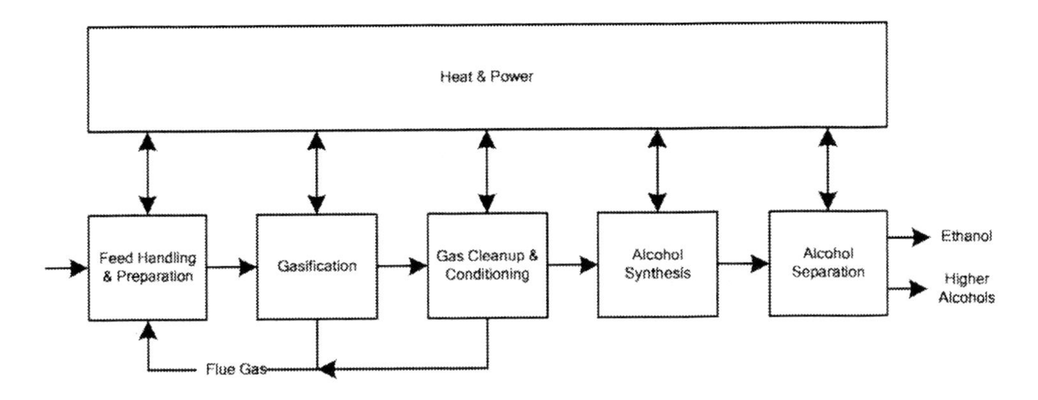

Figure 5. Block flow diagram

- *Heat & Power.* A conventional steam cycle produces heat (as steam) for the gasifier and reformer operations and electricity for internal power requirements (with the possibility of exporting excess electricity as a co-product). The steam cycle is integrated with the biomass conversion process. Pre-heaters, steam generators, and super-heaters are integrated within the process design to create the steam. The steam will run through turbines to drive compressors, generate electricity or be withdrawn at various pressure levels for injection into the process. The condensate will be sent back to the steam cycle, de-gassed, and combined with make-up water.

A cooling water system is also included in the Aspen Plus model to determine the requirements of each cooling water heat exchanger within the biomass conversion process as well as the requirements of the cooling tower.

Previous analyses of gasification processes have shown the importance of properly utilizing the heat from the high temperature streams. A pinch analysis was performed to analyze the energy network of this ethanol production process. The pinch concept offers a systematic approach to optimize the energy integration of the process. Details of the pinch analysis will be discussed in Section 3.10.

2.3. Feedstock and Plant Size

Based upon expected availability per the Billion Ton Vision [6] study, the forest resources were chosen for the primary feedstock. The Billion Ton Vision study addressed short and long term availability issues for biomass feedstocks without giving specific time frames. The amounts are depicted in Figure 6. The upper sets of numbers (labeled "High Yield Growth With Energy Crops" and "High Yield Growth Without Energy Crops") are projections of availability that will depend upon changes to agricultural practices and the creation of a new energy crop industry. In the target year of 2012 it is most probable that the amounts labeled "Existing & Unexploited Resources" will be the only ones that can be counted on to supply a thermochemical processing facility. Notice that the expected availability of forest resources is nearly the same as that of agricultural resources. Prior studies for biochemical processing have largely focused on using agricultural resources. It makes sense to base thermochemical processing on the forest resources. TC processing could fill an important need to provide a cost-effective technology to process this major portion of the expected biomass feedstock.

Past analyses have used hybrid poplar wood chips delivered at 50 wt% moisture to model forest resources [9]; the same will be done here. The ultimate analysis for the feed used in this study is given in Table 2. Performance and cost effects due to composition and moisture content were examined as part of the sensitivity analysis and alternate scenarios.

The design plant size for this study was chosen to match that of the Aden et al. biochemical process [3], 2,000 dry tonne/day (2,205 dry ton/day). With an expected 8,406 operating hours per year (96% operating factor) the annual feedstock requirement is 700,000 dry tonne/yr (772,000 dry ton/yr). As can be seen in Figure 6 this is a small portion of the 140 million dry ton/yr of forest resources potentially available. Cost effects due to plant size were examined as part of the sensitivity analysis.

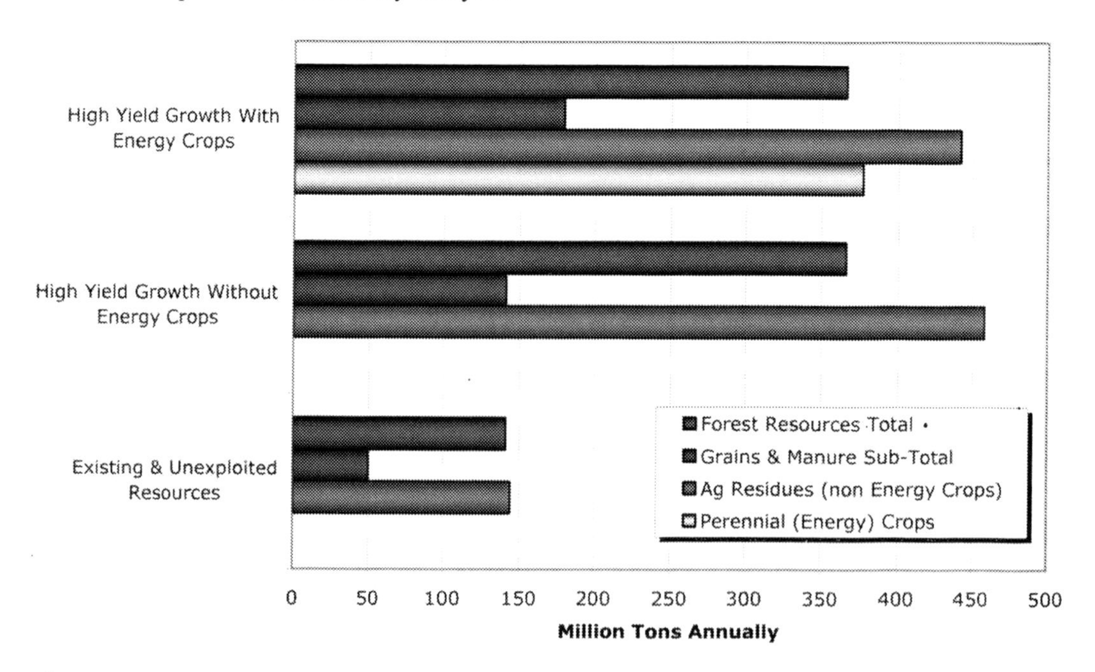

Figure 6. Expected availability of biomass

Table 2. Ultimate Analysis of Hybrid Poplar Feed

Component (wt%, dry basis29)	
Carbon	50.88
Hydrogen	6.04
Nitrogen	0.17
Sulfur	0.09
Oxygen	41.90
Ash	0.92
Heating value[c] (Btu/lb):	8,671 HHV[d] 8,060 LHV[e]

The delivered feedstock cost was chosen to match recent analyses done at Idaho National Laboratory (INL) [30] to target $35 per dry ton by 2012. Cost effects due to feedstock cost were also examined as part of the sensitivity analysis.

3. Process Design

3.1. Process Design Basis

The process design developed for this study is based upon the current operation and R&D performance goals for the catalytic tar destruction and heteroatom removal work at NREL and alcohol synthesis work at NREL and PNNL. This target design shows the effect of meeting these specific research and development (R&D) goals.

The process broadly consists of the following sections:

- Feed handling and drying
- Gasification
- Gas clean up and conditioning
- Alcohol synthesis
- Alcohol separation
- Integrated steam system and power generation cycle
- Cooling water and other utilities

3.2. Feed Handling and Drying – Area 100

This section of the process accommodates the delivery of biomass feedstock, short term storage on-site, and the preparation of the feedstock for processing in the gasifier. The design is based upon a woody feedstock. It is expected that a feed handling area for agricultural residues or energy crops would be very similar.

The feed handling and drying section are shown in PFD-P800-A101 and PFD-P800-A102. Wood chips are delivered to the plant primarily via trucks. However, it is envisioned that there could be some train transport. Assuming that each truck capacity is about 25 tons [31], this means that if the wood, at a moisture content of 50%, was delivered to the plant via truck transport only, then 176 truck deliveries per day would be required. As the trucks enter the plant they are weighed (M-101) and the wood chips are dumped into a storage pile. From the storage pile, the wood chips are conveyed (C-102) through a magnetic separator (S-101) and screened (S-102). Particles larger than 2 inches are sent through a hammer mill (T-102/M-102) for further size reduction. Front end loaders transfer the wood chips to the dryer feed bins (T-103).

Drying is accomplished by direct contact of the biomass feed with hot flue gas. Because of the large plant size there are two identical, parallel feed handling and drying trains. The wet wood chips enter each rotary biomass dryer (M-104) through a dryer feed screw conveyor (C-104). The wood is dried to a moisture content of 5 wt% with flue gas from the char combustor (R-202) and tar reformer's fuel combustor (R-301). The exhaust gas exiting the dryer is sent through a cyclone (S-103) and baghouse filter (S-104) to remove particulates prior to being emitted to the atmosphere. The stack temperature of the flue gas is set at 62° above the dew point of the gas, 235°F (113°C). The stack temperature is controlled by cooling the hot flue gas from the char combustor and the tar reformer with two steam boilers (H-286B and H-31 1B) prior to entering the dryer. This generated steam is added to the common steam drum (T-604) (see section on Steam System and Power Generation – Area 600). The dried biomass is then conveyed to the gasifier train (T-104/C-105).

Equipment costs were derived from the biochemical design report that utilized poplar as a feedstock. [5]

3.3. Gasification – Area 200

This section of the process converts a mixture of dry feedstock and steam to syngas and char. Heat is provided in an indirect manner — by circulating olivine that is heated by the combustion of the char downstream of the gasifier. The steam primarily acts as a fluidizing medium in the gasifier and also participates in certain reactions when high gasifier temperatures are reached.

From the feed handling and drying section, the dried wood enters the gasifier section as shown in PFD-P800-A201. Because of the plant size, it is assumed that there are two parallel gasifier trains. The gasifier (R-20 1) used in this analysis is a low-pressure indirectly-heated circulating fluidized bed (CFB) gasifier. The gasifier was modeled using correlations based on run data from the Battelle Columbus Laboratory (BCL) 9 tonne/day test facility (see Appendix I).

Heat for the endothermic gasification reactions is supplied by circulating a hot medium between the gasifier vessel and the char combustor (R-202). In this case the medium is synthetic olivine, a calcined magnesium silicate, primarily Enstatite ($MgSiO_3$), Forsterite (Mg_2SiO_3), and Hematite (Fe2O3), used as a heat transfer solid for various applications. A small amount of MgO must be added to the fresh olivine to avoid the formation of glass-like bed agglomerations that would result from the biomass potassium interacting with the silicate compounds. The MgO titrates the potassium in the feed ash. Without MgO addition, the potassium will form glass, K2SiO4, with the silica in the system. $K2SiO_4$ has a low melting point (~930°F, 500°C) and its formation will cause the bed media to become sticky, agglomerate, and eventually defluidize. Adding MgO makes the potassium form a high melting (~2,370°F, 1,300°C) ternary eutectic with the silica, thus sequestering it. Potassium carry-over in the gasifier/combustor cyclones is also significantly reduced. The ash content of the feed is assumed to contain 0.2 wt% potassium. The MgO flow rate is set at two times the molar flow rate of potassium.

The gasifier fluidization medium is steam that is supplied from the steam cycle (Steam System and Power Generation – Area 600). The steam-to-feed ratio is 0.4 lb of steam/lb of bone dry biomass. The gasifier pressure is 23 psia. The olivine circulating flow rate is 27 lb of olivine/lb of bone dry wood. Fresh olivine is added at a rate of 0.01% of the circulating rate to account for losses. The char combustor is operated with 20% excess air.

Both the gasifier and the char combustor temperatures are allowed to "float" and are dictated from the energy balances around the gasifier and combustor. In general, the more char created, the higher the char combustor temperature; but the higher the char combustor temperature, the higher the resulting gasifier temperature, resulting in less char. In this way the gasifier and char combustor temperatures tend to find an equilibrium position. For the design case the resulting gasifier temperature is 1,633°F (889°C) and the char combustor is 1,823°F (995°C). The composition of the outlet gas from the gasifier is shown in Table 3.

Particulate removal from the raw syngas exiting the gasifier is performed using two-stage cyclone separators. Nearly all of the olivine and char (99.9% of both) is separated in the primary gasifier cyclone (S-201) and gravity-fed to the char combustor. A secondary cyclone (S-202) removes 90% of any residual fines. The char that is formed in the gasifier is burned in the combustor to reheat the olivine. The primary combustor cyclone (S-203) separates the olivine (99.9%) from the combustion gases and the olivine is gravity-fed back to the gasifier. Ash and any sand particles that are carried over in the flue gas exiting the combustor are removed in the secondary combustor cyclone (99.9% separation in S-204) followed by an electrostatic precipitator (S-205) which removes the remaining residual amount of solid particles. The sand and ash mixture from the secondary flue gas cyclone and precipitator are land filled but prior to this the solids are cooled and water is added to the sand/ash stream for conditioning to prevent the mixture from being too dusty to handle. First the ash and sand mixture is cooled to 3 00°F (149°C) using the water cooled screw conveyor (M-20 1) then water is added directly to the mixture until the mixture water content is 10 wt%.

Capital costs for the equipment in this section are described in detail in Section 3 of this report. The operating costs for this section are listed in Appendix E and consist of makeup MgO and olivine, and sand/ash removal.

Table 3. Gasifier Operating Parameters, Gas Compositions, and Efficiencies

Gasifier Variable	Value	
Temperature	1,633°F (890°C)	
Pressure	23 psia (1.6 bar)	
Gasifier outlet gas composition	mol% (wet)	mol% (dry)
H_2	15.0	25.1
CO_2	7.4	12.4
CO	25.1	41.9
H_2O	40.2	--
CH_4	9.0	15.1
C_2H_2	0.3	0.4
C_2H_4	2.5	4.1
C_2H_6	0.1	0.2
C_6H_6	0.1	0.1
tar ($C_{10}H_8$)	0.1	0.2
NH_3	0.2	0.3
H_2S	0.04	0.07
H_2:CO molar ratio	0.60	
Gasifier Efficiency	76.6% HHV basis 76.1% LHV basis	

Table 4. Current and Target Design Performance of Tar Reformer

Compound	Experimental Conversion to CO & H_2	Target Conversion to CO & H_2
Methane (CH_4)	20%	80%
Ethane (C_2H_6)	90%	99%
Ethylene (C_2H_4)	50%	90%
Tars (C_{10+})	95%	99.9%
Benzene (C_6H_6)	70%	99%
Ammonia (NH_3)[f]	70%	90%

3.4. Gas Cleanup and Conditioning – Area 300

This section of the process cleans up and conditions the syngas so that the gas can be synthesized into alcohol. The type and the extent of cleanup are dictated by the requirements of the synthesis catalyst:

- The tars in the syngas are reformed to additional CO and H_2.
- Particulates are removed by quenching.
- Acid gases (CO_2 and H_2S) are removed.
- The syngas is compressed.

The gas from the secondary gasifier cyclone is sent to the catalytic tar reformer (R-303). In this bubbling fluidized bed reactor the hydrocarbons are converted to CO and H_2 while NH3 is converted to N_2 and H_2. In the Aspen simulation, the conversion of each compound is set to match targets that are believed to be attainable through near-term research efforts. Table 4 gives the current experimental conversions (for deactivated catalyst) that have been achieved at NREL [32] and the conversions used in the simulation corresponding to the 2012 research targets.

In the Aspen simulation the tar reformer operates isothermally at 1,633°F. An implicit assumption in this mode of operation is that the energy needed for the endothermic reforming reactions can be transferred into the catalyst bed. Although conceptual reactor designs are readily created for providing the heat of reaction from the fuel combustion area directly into the reformer catalyst bed, in practice this may be a difficult and prohibitively expensive design option requiring internal heat transfer tubes operating at high temperatures. An alternate approach, not used in this study, would be to preheat the process gas upstream of the reformer above the current reformer exit temperature, and operate the reformer adiabatically with a resulting temperature drop across the bed and a lower exit gas temperature. In this configuration, the required inlet and exit gas temperatures would be set by the extent of conversion, the kinetics of the reforming reactions, and the amount of catalyst in the reactor.

The composition of the gas from the tar reformer can be seen in Table 5.

Prior to the quench step, the hot syngas is cooled to 300°F (149°C) with heat exchangers (H301A-C) that are integrated in the steam cycle (see section Steam System and Power Generation – Area 600). After this direct cooling of the syngas, additional cooling is carried out via water scrubbing (M-302 and M-301), shown in PFD-P800-A302. The scrubber also removes impurities such as particulates, residual ammonia, and any residual tars. The scrubbing system consists of a venturi scrubber (M-302) and quench chamber (M-301). The scrubbing system quench water is a closed recirculation loop with heat rejected to the cooling tower and a continuous blow down rate of approximately 2.3 gallons per minute (gpm) that is sent to a waste water treatment facility. The quench water flow rate is determined by adjusting its circulation rate until its exit temperature from the quench water recirculation cooler (H-301) is 110°F (43°C). Any solids that settle out in T-301 are sent off-site for treatment as well. For modeling purposes, the water content of the sludge stream was set at 50 wt%.

The quench step cools the syngas to a temperature of 140°F (60°C). The syngas is then compressed using a five-stage centrifugal compressor with interstage cooling as shown in PFDP800-A303. The compressor was modeled such that each section has a polytropic efficiency of 78% and intercooler outlet temperatures of 140°F (60°C). The interstage coolers are forced air heat exchangers.

Depending on the specific catalysts being used downstream of the tar reformer, varying concentrations of acid gas compounds can be tolerated in the syngas. For example, sulfur concentrations as H_2S are required to be below 0.1 ppm for copper based synthesis catalysts. This design is based upon sulfided molybdenum catalysts which actually require up to 100 ppm of H_2S in the syngas to maintain catalyst activity. Because the syngas exiting the gasifier contains almost 400 ppmv of H_2S, some level of sulfur removal will be required by any of the synthesis catalysts currently of interest.

Carbon dioxide is the other acid gas that needs to be removed in the syngas conditioning process. Similar to the sulfur compounds, the acceptable level of CO_2 depends on the specific catalyst being used in the synthesis reactor to make alcohols. Some synthesis catalysts require low levels of CO_2 while others, such as the sulfided molybdenum catalysts can tolerate relatively high CO_2 levels compared to the sulfur species. CO_2 is a major component of the gasification product, so significant amounts of CO_2 may need to be removed upstream of the synthesis reactor.

Since the catalyst selected for this study is a sulfided catalyst that is tolerant of sulfur up to 100 ppmv and CO_2 up to 7 mol% (see Appendix J for more detail), a design that can provide for the removal of both sulfur and carbon dioxide was chosen. An amine system capable of selectively removing CO_2 and H2S from the main process syngas stream is used. The amine assumed for this study is monoethanol amine (MEA), based on the recommendation by Nexant [33].

The acid gas scrubber was simulated using a simplified model of SEP blocks and specifying the amount of CO_2 and H_2S needing to be removed to meet design specifications of 50 ppmv H_2S and 5 mol% CO_2 at the synthesis reactor inlet, including any recycle streams to that unit operation. The amine system heating and cooling duties were calculated using information taken from section 21 of the GPSA Data Handbook [34]. This method gave a heat duty of 2660 Btu per pound of CO_2 removed, with a similar magnitude cooling duty provided by forced-air cooling fans. Power requirements for pumping and fans were also calculated using GPSA recommended values. The acid gas scrubber operating values for the base case are given below.

Table 4B. Acid Gas Scrubber Operating Values for Base Case

Acid Gas Removal Parameter	Value
Amine Used	Monoethanol amine (MEA)
Amine Concentration	35 wt%
Amine Circ. Rate	1,945 gpm
Amine Temp. @ Absorber	110°F
Absorber Pressure	450 psia
Stripper Condenser Temperature	212°F
Stripper Reboiler Temperature	237°F
Stripper Pressure	65 psia
Stripper Reboiler Duty	140.1 MMBTU/hr
Stripper Condenser Duty	93.4 MMBTU/hr
Amine Cooler Duty	46.7 MMBTU/hr
Heat Duty per Pound CO_2 removed	2,660 Btu/lb

If a highly CO_2 -tolerant alcohol synthesis catalyst is used, it may become possible to use other syngas conditioning processes or methods to selectively remove H_2S, with less energy and possibly at a significantly lower capital cost.

The acid gases removed in the amine scrubber are then stripped to regenerate the sorbent and sent through a sulfur removal operation using a liquid phase oxidation process as shown in PFDP800-A305. The combined Amine/ LO-CAT process will remove the sulfur and CO_2 to the levels desired for the selected molysulfide catalyst [35]. Although, there are several liquid-phase oxidation processes for H_2S removal and conversion available today, the LO-CAT process was selected because of its progress in minimizing catalyst degradation and its environmentally- benign catalyst. LO-CAT is an iron chelate-based process that consists of a venturi precontactor (M-303), liquid-filled absorber (M-304), air-blown oxidizer (R-301), air blower (K-302), solution circulation pump (P-303) and solution cooler (H-305). Elemental sulfur is produced in the oxidizer and, since there is such a small amount (1.3 ton/day), it is stockpiled either for eventual disposal or sold as an unconditioned product. The LO-CAT process was modeled to remove the H_2S to a concentration of 10 ppmv in the CO_2 vent effluent from the amine scrubber. The air flow rate for re-oxidizing the LO-CAT solution was included in the simulation and calculated based on the requirement of 2 moles of O_2 per mole of H2S. Prior to entering the LO- CAT system the gas stream is superheated to 10°F (5.6°C) above its dew point in preheater (H304), which in this process is equivalent to 120°F. This degree of superheating is required for the LO-CAT system. The CO_2 from the LO-CAT unit is vented to the atmosphere.

The capital costs for the equipment in this section are described in further detail in the Appendices. The operating costs consist of makeup reforming catalyst, LO-CAT and amine chemical makeup, as well as reforming catalyst disposal cost and WWT. These are described in further detail in Section 3.

3.5. Alcohol Synthesis – Area 400

The alcohol synthesis reactor system is the heart of the entire process. Entering this process area, the syngas has been reformed, quenched, compressed and treated to have acid gas concentrations (H_2S, CO_2) reduced. After that, it is further compressed and heated to the synthesis reaction conditions of 1,000 psia and 570°F (300°C). The syngas is converted to the alcohol mixture across a fixed bed catalyst. The product gas is subsequently cooled, allowing the alcohols to condense and separate from the unconverted syngas. The liquid alcohols are then sent to alcohol separation and purification (Area 500). The residual gas stream is recycled back to the tar reformer with a small purge to fuel combustion (5%).

Research on alcohol synthesis catalysts has waxed and waned over many decades for a variety of reasons. In order to review the status of mixed alcohol technology and how it has developed over the past 20 years, two activities were initiated. First, a literature search was conducted. This search and its findings are described in more detail in Appendix J, along with a discussion on specific terminology, such as "yield", "selectivity", and "conversion". These terms will be used throughout the remainder of this document. Second, an engineering consulting company (Nexant) was hired to document the current state of technology with regards to mixed alcohols production and higher alcohol synthesis. Their results are published in an NREL subcontract [36] report.

Table 5. Target Design Tar Reformer Conditions and Outlet Gas Composition

Tar Reformer Variable	Value	
Tar reformer inlet temperature	1,633°F (890°C)	
Tar reformer outlet temperature	1,633°F (890°C)	
Tar reformer outlet gas composition	mol% (wet)	mol% (dry)
H_2	37.4	43.0
CO_2	9.9	11.4
CO	37.4	43.0
H_2O	13.0	---
CH_4	1.2	1.4
C_2H_2	0.01	0.01
C_2H_4	0.11	0.13
C_2H_6	10.8 ppmv	12.4 ppmv
C_6H_6	2.7 ppmv	3.1 ppmv
tar ($C_{10}H_8$)	0.5 ppmv	0.6 ppmv
NH_3	0.01	0.01
H_2S	0.02	0.02
N_2	0.72	0.83
H_2:CO molar ratio	1.00	

Based on the results of this background technology evaluation, a modified Fischer-Tropsch catalyst was used for this process design, specifically a molybdenum-disulfide-based (MoS_2) catalyst. The former Dow/UCC catalyst was chosen as the basis because of its relatively high ethanol selectivity and because its product slate is a mixture of linear alcohols (as opposed to the branched alcohols that result from modified methanol catalysts). This particular catalyst uses high surface area MoS_2 promoted with alkali metal salts (e.g.

potassium carbonate) and cobalt (CoS). These promoters shift the product slate from hydrocarbons to alcohols, and can either be supported on alumina or activated carbon, or be used unsupported.

Table 6. Process Conditions for Mixed Alcohols Synthesis

Parameter	"State of Technology" Conditions [41]	Target Conditions Used in Process Design & Aspen Model
Temperature (°C)	~ 300	300
Pressure (psia)	1500 - 2000	1000
H_2/CO ratio	1.0 – 1.2	1.0
CO_2 concentration (mol%)	0% - 7%	5.0%
Sulfur concentration (ppmv)	50 - 100	50

Table 6 lists several process and syngas conditioning requirements for this synthesis reaction. These include both experimentally verified conditions typical of those found in literature, as well as targeted conditions from the OBP-funded research plan used in the model.

Though the synthesis reactor is modeled as operating isothermally, it is recognized that maintaining a constant temperature in a fixed bed reactor system would be difficult, especially since these reactions are highly exothermic. Temperature has a significant impact on the alcohol selectivity and product distribution. High pressures are typically required to ensure the production of alcohols. MoS_2 catalysts are efficient Fischer-Tropsch (FT) catalysts at ambient or low pressures. However, significantly raising the pressure (in addition to promoting with alkali) helps to shift the pathways from hydrocarbon production towards alcohol production. However, compression requirements for achieving these pressures can be quite substantial. Thus, targeting a catalyst that achieves optimal performance at lower pressures can potentially provide significant cost savings.

The CO_2 concentration requirements for the syngas are less well-known. Herman [37] states that in the first Dow patent application, the presence of larger amounts of CO_2 in the synthesis gas retarded the catalyst activity. Further study showed that increasing the CO_2 concentration to 30 vol% decreased the CO conversion but did not significantly alter the alcohol:hydrocarbon ratio of the product. With CO_2 concentrations up to 6.7 vol%, the extent of CO conversion is not affected; however, higher chain alcohol yield relative to methanol does tend to decrease. This is why CO_2 concentrations were reduced to 5 mol% in the model using the amine system as part of syngas conditioning. The effect of CO_2 concentration on alcohol production will be studied in future laboratory experiments.

One of the benefits of this catalyst is its sulfur tolerance. It must be continuously sulfided to maintain its activity; thus an inlet gas concentration of 50 ppmv H_2S is maintained. Concentrations above 100 ppmv inhibit the reaction rate and higher alcohol selectivity.

The overall stoichiometric reaction for alcohol synthesis can be summarized as:

$$n\,CO + 2n\,H_2 \rightarrow C_nH_{2n+1}OH + (n-1)\,H_2O$$

Stoichiometry suggests an optimum H_2:CO ratio of 2.0. However this catalyst maintains significant water-gas shift activity and will generate its own H_2 from CO and H_2O:

$$CO + H_2O \rightarrow H_2 + CO_2.$$

This shifts the optimal ratio closer to 1.0 and also shifts the primary byproduct from water to CO_2. Experiments [38] have been typically conducted using ratios in the range of 1.0 to 1.2.

The compressor (K-4 10) in this area is a 3-stage steam-driven compressor that takes the syngas from 415 psia to 1000 psia, requiring 9,420 HP (assuming a polytropic efficiency of 78%). The outlet syngas from the compressor is then mixed with recycled methanol from Alcohol Purification (Area 500), heated to 570°F (300°C), and sent to the reactor. The capital cost for the compressor was developed using Questimate.

The mixed alcohol synthesis reactor is a fixed-bed reactor system that contains the MoS_2 catalyst. Because this is a net exothermic reaction system, water is cross exchanged with the reactor to produce steam for the process while helping to maintain a constant reactor temperature. Questimate was used to develop the reactor capital cost.

The purchase price of the catalyst itself was estimated at $5.25/lb based on conversations NREL researchers had with CRITERION [39], a petroleum/hydrocarbon catalyst provider. This represents a generalized cost of Molybdenum-based catalyst at around $5/lb being sulfided for an additional $0.25/lb. In addition, NREL was able to speak with Dow catalyst experts [40] who said that in today's market the raw material costs for producing such a catalyst system would run about $20/lb. Adding more cost for the catalyst preparation would bring that cost between $22-40/lb. However, these costs could go down as demand goes up, and quite substantially if it gets to large enough scale.

In reality, each company developing a process like this will have their own proprietary catalyst and associated formulation. The costs for these catalysts are difficult to predict at the present time since so few providers of mixed alcohols catalyst currently exist (and will likely be negotiated). Nexant also provided information on general catalyst metals price ranges in their report. They reported Molybdenum ranging from $2 – 40/lb.

The lifetime of the catalyst was assumed to be 5 years. While existing mixed alcohols catalysts have not been tested for this long, they have operated for over 8,000 hours (roughly 1 year of continuous operating time) with little or no loss in performance.

The reactor was modeled as a simple conversion-specified reactor using a series of alcohol and hydrocarbon production reactions as shown in Table 7. The propane, butane, and pentane+ reactions are set to zero because the catalyst will likely not favor these reactions. The specific conversions of each of the other reactions were set in order to reach catalyst performance targets, see Table 9. Those targets are shown in Table 8 along with values for those parameters typically found in literature.

The individual target values are less important than the net result of the entire collection. For example, a catalyst system can have a high CO conversion well above 40%, but if most of that CO is converted to methane or CO_2, then the alcohol selectivities would be very low and the entire process economics would suffer. Likewise, if the catalyst had a high CO conversion and selectivity, but had very low productivity, a much larger reactor would have to be built to accommodate the volume of catalyst required. The set of targets shown above are improvements over current literature values, but were chosen as targets believed to be achievable through catalyst research and development. There is precedent for these results from other catalyst systems. For example, FT catalysts are currently capable of CO conversions above 70% [42]. Also commercial methanol catalysts have productivities over 1000 g/kg-catalyst/hr [37].

Table 7. System of Reactions for Mixed Alcohol Synthesis

Water-Gas Shift	$CO + H_2O \leftrightarrow H_2 + CO_2$
Methanol	$CO + 2H_2 \rightarrow CH_3OH$
Methane	$CH_3OH + H_2 \rightarrow CH_4 + H_2O$
Ethanol	$CO + 2H_2 + CH_3OH \rightarrow C_2H_5OH + H_2O$
Ethane	$C_2H_5OH + H_2 \rightarrow$ C2H6 + H2O
Propanol	$CO + 2H_2 + C_2H_5OH \rightarrow C_3H_7OH + H_2O$
Propane	$C_3H_7OH + H_2 \rightarrow C_3H_8 + H_2O$
n-Butanol	$CO + 2H_2 + C_3H_7OH \rightarrow C_4H_9OH + H_2O$
Butane	$C_4H_9OH + H_2 \rightarrow C_4H_{10} + H_2O$
Pentanol+	$CO + 2H_2 + C_4H_9OH \rightarrow C_5H_{11}OH + H_2O$
Pentane+	$C_5H_{11}OH + H_2 \rightarrow C_5H_{12} + H_2O$

Table 8. Mixed Alcohol Reaction Performance Results

Result	"State of Technology" Value Ranges [37, 41]	Target Results Used in Process Design & Aspen Model
Total CO Conversion (per-pass)	10% - 40%	60%
Total Alcohol Selectivity (CO_2-free basis)	70% - 80%	90%
Gas Hourly Space Velocity (hr^{-1})	1600 – 12,000	4000
Catalyst Alcohol Productivity (g/kg- catalyst/hr)[g]	150 – 350	600

The reaction conversions were also set to achieve a certain product distribution of alcohols. The mixed alcohol products described in literature are often high in methanol, but contain a wide distribution of several different alcohols. The product distributions described by Dow and SRI are shown in Table 9 along with the relative product concentrations calculated by the model.

The most significant differences between the NREL model product distribution and those shown in literature are with regards to the methanol and ethanol distributions. This is primarily due to the almost complete recycle of methanol within this process. In the alcohol purification section downstream, virtually all methanol is recovered via distillation and recycled back to mix with the compressed syngas. This is done in order to increase the production of ethanol and higher alcohols. This concept has been proposed in literature, but data from testing in an integrated setting has not been seen. In literature, experiments are often conducted on closed or batch systems and do not examine the potential impacts of recycled compounds or other integration issues. However, this catalyst is known to have methanol decomposition functionality which indicates that methanol in the feed will not be detrimental to the reaction. The effects of recycled methanol will be examined experimentally as research progresses.

Table 9. Mixed Alcohol Product Distributions

Alcohol	Dow [43] (wt %)	SRI [44] (wt%)	NREL Model (wt%)*
Methanol	30-70%	30.77%	5.01%
Ethanol	34.5%	46.12%	70.66%
Propanol	7.7%	13.3%	10.07%
Butanol	1.4%	4.14%	1.25%
Pentanol +	1.5%	2.04%	0.17%
Acetates (C1 & C2)	2.5%	3.63%	
Others			10.98%
Water	2.4%		1.86%
Total	**100%**	**100%**	**100%**

* Prior to alcohol purification and methanol recycle

A kinetic model was used to guide these conversion assumptions to help predict how the catalyst may perform as a result of significant methanol recycle. Very few kinetic models have been developed for this catalyst system [45, 46, 47]. Of these, only Gunturu examined the possibility of methanol recycle. Therefore NREL reproduced this kinetic model using Polymath software. This kinetic model predicted that methanol entering the reactor would largely be converted to ethanol and methane. This model also predicts that maintaining high partial pressures of methanol in the reactor would further reduce the production of alcohols higher than ethanol. More detailed discussion on the kinetic model can be found in Appendix K.

After the reactor, the effluent is cooled to 110°F (43°C) through a series of heat exchangers while maintaining high pressure. First, the reacted syngas is cross exchanged with cooler process streams, lowering the temperature to 200°F (93°C). Air-cooled exchangers then bring the temperature down to 140°F (60°C). The final 30°F (17°C) drop is provided by cooling water. A knock-out drum (S-50 1) is then used to separate the liquids (primarily alcohols) from the remaining gas, which is comprised of unconverted syngas, CO_2, and methane. Aspen Plus contains other physical property packages that model non-ideal liquid systems much better than the Redlich-Kwong-Soave (RKS) equation of state used throughout the model. Therefore, the Non-Random Two-Liquid (NRTL) package was used to model the alcohol condensation.

From here, the liquid crude alcohols are sent to product purification while the residual syngas is superheated to 1500°F (816°C) and sent through an expander to generate additional power for the process. The pressure is dropped from 970 to 35 psia prior to being recycled to the tar reformer. A 5% purge stream is sent to fuel combustion.

Alternate configurations will be discussed later in this report as will the economic sensitivity of certain synthesis parameters. One particular variation would be to recycle the unconverted syngas to the throat of the synthesis reactor instead of to the tar reformer. This would save money on upstream equipment costs because of lower process throughput, but would also lower yields because the CO_2 would build up in the recycle loop. The limit to the amount of unconverted syngas that could be recycled to the reactor is less than 50% because this would cause the H_2:CO ratio to grow well above 1.2.

Future experiments and analysis will examine the impacts of methanol recycle, and of variations in concentration of CO_2, CH_4, and other compounds. Alternate reactor designs will also be examined. For example, FT technology largely has switched to slurry reactors instead of fixed-bed reactors because the slurry fluidization achieves better heat and mass transfer properties that allow, in turn, for higher conversions. Such improvements could help to achieve the conversion targets outlined above and reduce the costs of major equipment items.

3.6. Alcohol Separation – Area 500

The mixed alcohol stream from Area 400 is sent to Area 500 where it is de-gassed, dried, and separated into three streams: methanol, ethanol, and mixed higher-molecular weight alcohols. The methanol stream is used to back-flush the molecular sieve drying column and then recycled, along with the water removed during back flushing, to the inlet of the alcohol synthesis reactor in Area 400. The ethanol and mixed alcohol streams are cooled and sent to product storage tanks.

Carbon dioxide is readily absorbed in alcohol. Although the majority of the non-condensable gases leaving the synthesis reactor are removed in the separator vessel, S-50 1, a significant quantity of these gases remains in the alcohol stream, especially at the high system pressure. These gases are removed by depressurizing from 970 to 60 psia. Most of the dissolved gasses separate from the alcohols in the knock-out vessel S-502. This gas stream is made up primarily of carbon dioxide with some small amounts of hydrocarbons and alcohols; it is recycled to the Tar Reformer in Area 300. After being vaporized by cross, exchanging with steam to a 20°F (11°C) superheated temperature, the alcohol stream goes to the molecular sieve dehydrator unit operation.

The molecular sieve dehydrator design was based upon previous biochemical ethanol studies [5, 3] and assumed to have similar performance with mixed alcohols. In the biochemical ethanol cases, the molecular sieve is used to dry ethanol after it is distilled to the azeotropic concentration of ethanol and water (92.5 wt% ethanol). The adsorbed water is flushed from the molecular sieves with a portion of the dried ethanol and recycled to the rectification column. The water ultimately leaves out the bottom of the distillation column. In this thermochemical process, however, it was determined that drying the entire mixed alcohol stream before any other separation would be preferable. The adsorbed water is desorbed from the molecular sieves with a combination of depressurization and flushing with methanol. This methanol/water mixture is then recycled back to the Alcohol Synthesis section (A400).

The molecular sieve units require a superheated vapor. The liquid mixed alcohol stream is vaporized, superheated, and then fed to one of two parallel adsorption columns. The adsorption column preferentially removes water and a small amount of the alcohols. While one adsorption bed is adsorbing water, the other is regenerating. The water is desorbed from the bed during regeneration by applying a vacuum and flushing with dry methanol from D-505. This methanol/water mixture is recycled back to the Alcohol Synthesis section (A400). This methanol/water mixture is cooled to 140°F (60°C) using a forced air heat exchanger, and separated from any uncondensed vapor. The gaseous stream is recycled to the Tar Reformer and the condensate is pumped to 1,000 psia in P-514, and mixed with high-pressure syngas from compressor K-410 in Area 400 upstream of the synthesis reactor pre-heater.

Table 10. Plant Power Requirements

Plant Section	Power Requirement (kW)
Feed Handling & Drying	742
Gasification	3,392
Tar Reforming, Cleanup, & Conditioning	1,798
Mixed Alcohol Synthesis	119
Alcohol Separation and Purification	256
Steam System & Power Generation	431 required 7,994 generated
Cooling Water & Other Utilities	529
Miscellaneous	727
Total plant power requirement	7,994

The dry mixed alcohol stream leaving the mol sieve dehydrator enters into the first of two distillation columns, D-504. D-504 is a typical distillation column using trays, overhead condenser, and a reboiler. The methanol and ethanol are separated from the incoming stream with 99% of the incoming ethanol being recovered in the overhead stream along with essentially all incoming methanol. The D-504 bottom stream consists of 99% of the incoming propanol, 1% of the incoming ethanol, and all of the butanol and pentanol. The mixed alcohol bottoms is considered a co-product of the plant and is cooled and sent to storage. The methanol/ethanol overhead stream from D-504 goes to a second distillation column, D-505, for further processing.

D-505 separates the methanol from the binary methyl/ethyl alcohol mixture. The ethanol recovery in D-505 is 99% of the incoming ethanol and has a maximum methanol concentration of 0.5 mole percent to meet product specifications for fuel ethanol. The ethanol, which exits from the bottom of D-505 is cooled before being sent to product storage. The methanol and small quantity of ethanol exiting the overhead of column D-505 is used to flush the mol sieve column during its regeneration step as explained above. Currently, all of the methanol from D-505 is recycled through the mol sieve dehydrator and then to the synthesis reactor in Area 400.

3.7. Steam System and Power Generation – Area 600

This process design includes a steam cycle that produces steam by recovering heat from the hot process streams throughout the plant. Steam demands for the process include the gasifier, amine system reboiler, alcohol purification reboilers, and LO-CAT preheater. Of these, only the steam to the gasifier is directly injected into the process; the rest of the plant heat demands are provided by indirect heat exchange of process streams with the steam and have condensate return loops. Power for internal plant loads is produced from the steam cycle using an extraction steam turbine/generator (M-602). Power is also produced from the process expander (K-412), which takes the unconverted syngas from 965 psia to 35 psia before being recycled to the tar reformer. Steam is supplied to the gasifier from the low pressure turbine exhaust stage. The plant energy balance is managed to generate only the amount of electricity required by the plant. The steam system and power generation area is shown in PFD-P800-A601, -A602, and -A603 in Appendix H.

A condensate collection tank (T-601) gathers condensate from the syngas compressors and from the process reboilers along with the steam turbine condensate and make-up water. The total condensate stream is heated to the saturation temperature and sent to the deaerator (T-603) to degas any dissolved gases out of the water. The water from the deaerator is first pumped to a pressure of 930 psia and then pre-heated to its saturation (bubble point) temperature using a series of exchangers. The saturated steam is collected in the steam drum (T-604). To prevent solids build up, water must be periodically discharged from the steam drum. The blowdown rate is equal to 2% of water circulation rate. The saturated steam from the steam drum is superheated with another series of exchangers. The superheated steam temperature and pressure were set as a result of pinch analysis. Superheated steam enters the turbine at 900°F and 850 psia and is expanded to a pressure of 175 psia. The remaining steam then enters the low pressure turbine and is expanded to a pressure of 65 psia. Here a slipstream of steam is removed and sent to the gasifier and other exchangers. Finally, the steam enters a condensing turbine and is expanded to a pressure of 1.5 psia. The steam is condensed in the steam turbine condenser (H-601) and the condensate re-circulated back to the condensate collection tank.

The integration of the individual heat exchangers can only be seen in the PFDs included in the Appendices. To close the heat balance of the system, the Aspen Plus model increases or decreases the water flowrate through the steam cycle until the heat balance of the system is met.

This process design assumes that the two compressors in this process (K301, K410) are steam-driven. All other drives for pumps, fans, etc are electric motors. Additionally, an allowance of 0.7 MW of excess power is made to total power requirement to account for miscellaneous usage and general electric needs (lights, computers, etc). Table 10 contains the power requirement of the plant broken out into the different plant sections. Because syngas compression is steam driven, it is not a demand on the power system, which makes the total power requirement much less than it would be if compression demands were included. The plant power demands and power production were designed specifically to be nearly equal. Therefore, no excess power is being sold to or purchased from the grid. This plant was designed to be as energy self-sufficient as possible. This was accomplished by burning a portion of the "dirty" unreformed syngas in the fuel combustor (Section 300). While this does have a negative impact on the overall alcohol yields of the process, it does negate the purchase of natural gas or grid power.

3.8. Cooling Water and Other Utilities – Area 700

The cooling water system is shown on PFD-P800-A701. A mechanical draft cooling tower (M701) provides cooling water to several heat exchangers in the plant. The tower utilizes large fans to force air through circulated water. Heat is transferred from the water to the surrounding air by the transfer of sensible and latent heat. Cooling water is used in the following pieces of equipment:

- the sand/ash cooler (M-201) which cools the sand/ash mixture from the gasifier/combustor
- the quench water recirculation cooler (M-301) which cools the water used in the syngas quench step
- the water-cooled aftercooler (H-303) which follows the syngas compressor and cools the syngas after the last stage of compression
- the LO-CAT absorbent solution cooler (H-305) which cools the regenerated solution that circulates between the oxidizer and absorber vessels
- the reacted syngas cooler (H-414) which cools the gas in order to condense out the liquid alcohols
- the end product finishing coolers (H-591, H-593) for both the higher alcohols co-product and the primary ethanol product
- the blowdown water-cooled cooler (H-603) which cools the blowdown from the steam drum
- the steam turbine condenser (H-601) which condenses the steam exiting the steam turbine

Make-up water for the cooling tower is supplied at 14.7 psia and 60°F (16°C). Water losses include evaporation, drift (water entrained in the cooling tower exhaust air), and tower basin blowdown. Drift losses were estimated to be 0.2% of the water supply. Evaporation losses and blowdown were calculated based on information and equations in Perry, et al. [27]. The cooling water returns to the process at a supply pressure of 65 psia and temperature is 90°F (3 2°C). The cooling water return temperature is 110°F (43°C).

An instrument air system is included to provide compressed air for both service and instruments. The instrument air system is shown on PFD-P800-A701. The system consists of an air compressor (K-70 1), dryer (S-70 1) and receiver (T-70 1). The instrument air is delivered at a pressure of 115 psia, a moisture dew point of -40°F (-40°C), and is oil free.

Other miscellaneous items that are taken into account in the design include:

- a firewater storage tank (T-702) and pump (P-702)
- a diesel tank (T-703) and pump (P-703) to fuel the front loaders
- an olivine truck scale with dump (M-702) and an olivine lock hopper (T-705) as well as an MgO lock hopper (T-706)
- a hydrazine storage tank (T-707) and pump (P-705) for oxygen scavenging in the cooling water

This equipment is shown on PFD-P800-A702.

3.9. Additional Design Information

Table 11 contains some additional information used in the Aspen Plus model and production design.

- In the GPSA *Engineering Data Book* [48], see Table 11.4 for typical design values for dry bulb and wet bulb temperature by geography. Selected values would cover summertime conditions for most of lower 48 states.
- In Weast [49], see F-172 for composition of dry air. Nitrogen value adjusted slightly to force mole fraction closure using only N_2, O_2, Ar, and CO_2 as air components.
- In Perry, et al. [27], see psychrometric chart, Figure 12-2, for moisture content of air.

3.10. Pinch Analysis

A pinch analysis was performed to analyze the energy network of the biomass gasification to ethanol production process. The pinch technology concept offers a systematic approach to optimum energy integration of the process. First temperature and enthalpy data were gathered for the "hot" process streams (i.e., those that must be cooled), "cold" process streams (i.e., those that must be heated), and utility streams (such as steam, flue gas, and cooling water). The minimum approach temperature was set at 42.6°F. A temperature versus enthalpy graph (the "composite curve") was constructed for the hot and cold process streams. These two curves are shifted so that they touch at the pinch point. From this shifted graph, a grand composite curve is constructed which plots the enthalpy differences between the hot and cold composite curves as a function of temperature. The composite curve is shown in Figure 7. From this figure the heat exchanger network of the system was determined.

The total heating enthalpy equals the total cooling enthalpy because the Aspen model is designed to adjust the water flowrate through the steam cycle until the heat balance in the system is met. Because no outside utilities were used in this process, all heating and cooling duties are satisfied through process-process interchanges or process-steam interchanges. The minimum vertical distance between the curves is $\Delta T_{min,}$ which is theoretically the smallest approach needed in the exchange network. For this design, the pinch occurs at ~ H = 280,000,000 BTU/hr, and the upper and lower pinch temperatures are 570.0°F and 527.4°F, respectively, giving a ΔT_{min} of 42.6°F.

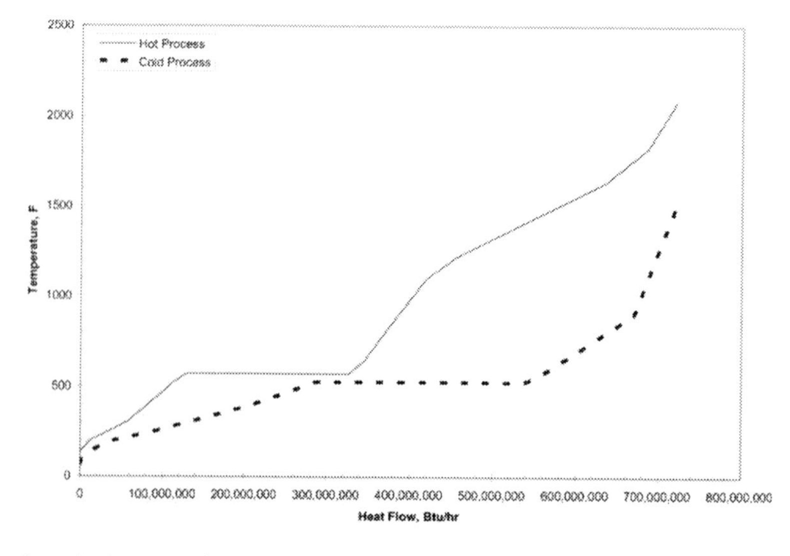

Figure 7. Pinch analysis composite curve

Table 11. Utility and Miscellaneous Design Information

Item	Design Information
Ambient air conditions (1,2, and 3)	Pressure: 14.7 psia $T_{Dry\ Bulb}$: 90°F $T_{Wet\ Bulb}$: 80°F Composition (mol%): N_2: 75.7% O_2: 20.3% Ar: 0.9% CO_2: 0.03% H_2O: 3.1%
Pressure drop allowance	Syngas compressor intercoolers = 2 psi Heat exchangers and packed beds = 5 psi

Design of the heat exchange network for the above the pinch and below pinch regions are done separately. While pinch theory teaches that multiple solutions are possible, this particular solution has the advantage that heat released by the alcohol synthesis reactor is dissipated by raising steam. This is a standard design practice for removing heat from methanol synthesis and other similar reactors. The left-hand side of the composite curve shows the below pinch curves are constrained at the pinch and are also nearly pinched at the very left-hand side in the ~ 100°F range. This makes heat exchanger network design below the pinch more difficult.

3.11. Energy Balance

Energy integration is extremely important to the overall economics and efficiency of this process. Therefore a detailed understanding of how and where the energy is utilized and recovered is required. Detailed energy balances around the major process areas were derived using data from the Aspen Plus simulation. Comparing the process energy inputs and outputs enables the energy efficiency of the process to be quantified. Also, tracing energy transfer between process areas makes it possible to identify areas of potential improvement to the energy efficiency.

The philosophy of defining the "energy potential" of a stream is somewhat different from what was done for the biochemical ethanol process design report [50]. For that analysis the definition of the energy potential was based upon the higher heating values (HHVs) of each component. This HHV basis is convenient when a process is primarily made up of aqueous streams in the liquid phase. Since liquid water at the standard temperature has a zero HHV, the contribution for any liquid water is very small, especially as compared to any other combustible material also present in the stream. However, the thermochemical ethanol production process differs significantly in that most of the process streams are in the gas phase. To remove the background contributions of the water, the energy potential is based instead upon the lower heating values (LHVs) of each component.

The total energy potential for a stream has other contributions beyond that of the heating value. Other energy contributions are:

- Sensible heat effect – the stream is at a temperature (and pressure) different from that of the standard conditions at which the heating values are defined.

- Latent heat effect – one or more components in the stream are in a different phase from that at which their heating values are defined.
- Non-ideal mixing effect – any heating or cooling due to blending dissimilar components in a mixture.

The procedure for actually calculating the energy potential of a stream is also different from what was done prior. When the biochemical ethanol process was analyzed, the contributions for the HHVs, the sensible heating effects, and the latent heat effects were directly computed and combined. The calculations of the sensible and latent heat effects were done in an approximate manner. For example, the sensible heat effect was estimated from the heat capacity at the stream's temperature, pressure, and composition; it was assumed that this heat capacity remained constant over the temperature range between the stream's temperature and the standard temperature. For the relatively low temperatures of the biochemical ethanol process systems, this assumption makes sense. However, for this thermochemical process design, this assumption is not accurate because of the much larger differences between the process stream temperatures and the standard temperature

The enthalpy values reported by Aspen Plus can actually be adjusted in a fairly simple manner to reflect either an HHV or LHV basis for the energy potential. The enthalpies calculated and reported by Aspen Plus are actually based upon a heat of formation for the energy potential of a stream. So, the reported enthalpies already include the sensible, latent, and non-ideal mixing effects. If certain constants in Aspen's enthalpy expressions could be modified to be based on either the components' HHVs or LHVs instead of the heats of formation then Aspen Plus would report the desired energy potential values. However, since the constants cannot be easily changed, the reported enthalpy values were adjusted instead as part of a spreadsheet calculation.

The factors used to adjust the reported enthalpies were calculated from the difference between each component's heat of combustion (LHV) and the reported pure component enthalpy at combustion conditions.

This process for thermochemical conversion of cellulosic biomass was designed with the goal of being as energy self-sufficient. Natural gas inputs that could be used to fire the char combustor and fuel combustor have been eliminated. Instead, a slipstream of "dirty" unreformed syngas is used to meet the fuel demand. The downside to this is a decrease in ethanol yield. In addition, the process was designed to require no electricity be purchased from the grid. Instead, the integrated combined heat and power system supplies all steam and electricity needed by the plant. Consequently no electricity is sold as a co-product either. The only saleable products are the fuel ethanol and a higher molecular weight mixed alcohol co-product.

The major process energy inputs and outlets are listed in Table 12, along with their energy flowrates. Each input and output is also ratioed to the biomass energy entering the system. The biomass is of course the primary energy input, however other energy inputs are required. Air is required for both the fuel combustor as well as the char combustor; however it remains a minor energy input. Some water is used to wet the ash leaving the gasification system, however, the majority of process water is used for boiler feed water makeup and cooling water makeup. A large negative energy flow value is associated with this because it enters the process as a liquid.

Table 12. Overall Energy Analysis (LHV basis)

	Energy Flow (MMBTU/hr, LHV basis)	Ratio to Feedstock Energy Flow
Energy Inlets		
Wood Chip Feedstock (wet)	1269.7	1.000
Natural Gas	0.0	0.000
Air	2.3	0.002
Olivine	0.0	0.000
MgO	0.0	0.000
Water	-133.4	-0.105
Tar Reforming Catalyst	0.0	0.000
Other	0.0	0.000
Total	**1138.6**	**0.897**
Energy Outlets		
Ethanol	619.1	0.488
Higher Alcohols Co-product	122.1	0.096
Cooling Tower Evaporation	17.0	0.013
Flue Gas	46.2	0.036
Sulfur	0.4	0.000
Compressor Heat	178.3	0.140
Heat from Air-cooled Exchangers	222.0	0.175
Vents to Atmosphere (including excess CO_2)	0.8	0.001
Sand and Ash	16.4	0.013
Catalyst Purge	0.0	0.000
Wastewater	-1.2	-0.001
Other	-82.5	-0.064
Total	**1138.6**	**0.897**

Besides the saleable alcohol products, other important process energy outlets also exist. There are two sources of flue gas: the char combustor and the reformer fuel combustor. Together, they total about 4% of the energy in the raw biomass. Cooling tower evaporative losses, excess CO_2 vent to the atmosphere, wastewater, and ash streams are also minor process energy outlets. However, two of the larger energy outlets come from air-cooled interstage cooling of the compressors, and from several other air-cooled heat exchangers. Together, these two loss categories represent over 30% of the energy that is not recovered within the process. The "other" category consists primarily of other losses from the cooling tower system (drift and blowdown), but also accounts for energy losses due to ambient heating effects and mechanical work (pump, compressor) efficiency losses.

Some of this lost heat could potentially be recovered by using cooling water instead of air-cooled exchangers. However, this would require additional makeup water, and limiting water usage throughout the process was a primary design consideration. Additional heat integration with process streams could also be examined, however, there comes a point where this becomes too complex and costly for a cost-effective design and practical operation.

Overall, the TC process is approximately 46% efficient on an LHV basis for moisture-free biomass, as shown in the Appendices. Table 12 shows that approximately 58% of the energy in the wet raw biomass is recovered in the two alcohol products. Improvements in these energy efficiencies could potentially result in additional cost savings to the process.

3.12. Water Issues

Water is required as a reactant, a fluidizing agent, and a cooling medium in this process. As a reactant, it participates in reforming and water gas shift reactions. Using the BCL gasifier, it also acts as the fluidizing agent in the form of steam. Its cooling uses are outlined in Section 3.8.

Water usage is becoming an increasingly important aspect of plant design, specifically with regards to today's ethanol plants. Most ethanol plants reside in the Midwest where many places are experiencing significant water supply concerns 51. For several years, significant areas of water stress have been reported during the growing season, while livestock and irrigation operations compete for the available resources.

Today's dry mill ethanol plants have a high degree of water recycle. In fact many plants use what is known as a "zero discharge" design where no process water is discharged to wastewater treatment. The use of centrifuges and evaporators enables this recycle of process water. Therefore, much of the consumptive water demand of an ethanol plant comes from the evaporative losses from the cooling tower and utility systems. Oftentimes well water is used to supply the water demands of the ethanol plants, which draws from the local aquifers that are not readily recharged. This is driven by the need for high quality water in the boiler system. Studies have shown that water usage by today's corn ethanol plants range from 3-7 gallons per gallon of ethanol produced. This means that a 50 MM gal/yr dry mill will use between 150-350 MM gallons/yr of water that is essentially a non-renewable resource. This ratio however has decreased over time from an average of 5.8 gal/gal in 1998 to 4.2 gal/gal in 2005.

Therefore, a primary design consideration for this process was the minimization of fresh water requirements, which therefore meant minimizing the cooling water demands and recycling process water as much as possible. Air-cooling was used in several areas of the process in place of cooling water (e.g. distillation condensers, compressor interstage cooling, etc). However there are some instances where cooling water is required to reach a sufficiently low temperature that air-cooling can not reach.

Table 13 quantifies the particular water demands of this design. Roughly 71% of the fresh water demand is from cooling tower makeup, with most of the remainder needed as makeup boiler feed water. Some of this water is directly injected into the gasifier, but other system losses (blowdown) also exist. The overall water demand is considerably less than today's ethanol plants. This design requires less than 2 gallons of fresh water for each gallon of ethanol produced. It may be worthwhile for the entire ethanol industry to more thoroughly investigate efficiency gains that are possible within these utility systems.

Table 13. Process Water Demands for Thermochemical Ethanol

Fresh Water Demands	lb per hour
Cooling Tower Makeup	84,672
Boiler Feed Makeup	34,176
Sand/ash Wetting	243
Total	**119,091**
Overall Water Demand (gal water / gal ethanol)	**1.94**

4. PROCESS ECONOMICS

The total project investment (based on total equipment cost) as well as variable and fixed operating costs is developed first. With these costs, a discounted cash flow analysis was used to determine the production cost of ethanol when the net present value of the project is zero. This section describes the cost areas and the assumptions made to complete the discounted cash flow analysis.

4.1. Capital Costs

The following sections discuss the methods and sources for determining the capital cost of each piece of equipment within the plant. A summary of the individual equipment costs can be found in Appendix D.

The capital cost estimates are based as much as possible on the design work done by Spath et al . for the hydrogen design report [9] and Aden et al. for the biochemical conversion design report [3]. The majority of the Spath et al. costs came from literature and Questimate (an equipment capital cost estimating software tool by Aspen Tech), not from vendor quotes. For these estimated costs, the purchased cost of the equipment was calculated and then cost factors were used to determine the installed equipment cost. This method of cost estimation has an expected accuracy of roughly +30% to -10%. The factors used in determining the total installed cost (TIC) of each piece of equipment are shown in Table 14 [52]. The Aden et al. cost estimates came from a variety of sources (including vendor quotes); the installation factors for these estimates may be significantly different from what is in Table 14.

The indirect costs (non-manufacturing fixed-capital investment costs) were also estimated as per Spath et al. using cost factors. The factors are shown in Table 15 [52] and have been put as percentages in terms of total purchased equipment cost, total installed cost (TIC), and total project investment (TPI, the sum of the TIC and the total indirect costs).

The biomass handling and drying costs as well as the gasification and gas clean up costs were estimated by Spath et al. using several reports by others that documented detailed design and cost estimates. Some of the reports gave costs for individual pieces of equipment while others lumped the equipment costs into areas. The costs from the reports were amalgamated into:

- feedstock handling and drying.
- gasification and clean up.

Table 14. General Cost Factors in Determining Total Installed Equipment Costs

	% of TPEC
Total Purchased Equipment Cost (TPEC)	100
Purchased equipment installation	39
Instrumentation and controls	26
Piping	31
Electrical systems	10
Buildings (including services)	29
Yard improvements	12
Total Installed Cost (TIC)	247

Table 15. Cost Factors for Indirect Costs

Indirect Costs	% of TPEC	% of TIC	% of TPI
Engineering	32	13	9
Construction	34	14	10
Legal and contractors fees	23	9	7
Project contingency	7.4	3	2
Total Indirect Costs	96.4	39	28

Costs from those reports scaled to a 2,000 bone dry tonne/day plant are given in Table 16. Table 17 gives the basic dryer and gasifier design basis for the references. Spath et al. used an average feed handling and drying cost from all of the literature sources and an average gasifier and gas clean up cost for the references using the BCL gasifier.

In this report, we have further broken apart the gasification and clean up costs into their respective areas. Based upon the Utrecht report [19] these were split 50/50 between the two areas.

The cost of reactors, heat exchangers, compressors, blowers and pumps were estimated for a "base" size using Questimate and then scaled using material and energy balance results from the Aspen Plus simulation. The reactors were sized based on a gas hourly space velocity (GHSV), where GHSV is measured at standard temperature and pressure, 60°F and 1 atm [61], and a height to diameter ratio of 2. The GHSV for the mixed alcohol reactor and tar reformer were set at 4,000/hr and 2475/hr, respectively. These are in agreement with typical values given by Kohl and Nielsen [62]. The heat exchanger costs were mostly developed based on the required surface area as calculated from the heat transfer equation appropriate for a 1-1 shell and tube heat exchanger:

$$Q = UA(\Delta T)_{lm} \quad \Rightarrow \quad A = \frac{Q}{U(\Delta T)_{lm}}$$

where Q is the heat duty, U is the heat transfer coefficient, A is the exchanger surface area, and $(\Delta T)_{lm}$ is the log mean temperature difference. The heat transfer coefficients were estimated from literature sources, primarily Perry, et al [27]. However, many of the exchangers used in the pinch analysis are subsequently scaled from their calculated duties. At present, these duties will not change as the process changes, unless the pinch calculations are specifically updated. This is acceptable as long as the total cost of the heat exchange network remains a small fraction of the overall minimum ethanol plant gate price, and as long as plant scale does not change significantly.

Table 16. Feed Handling & Drying and Gasifier & Gas Clean Up Costs from the Literature Scaled to 2,000 tonne/day plant

Reference	Scaled Feed Handling and Drying Cost $K (2002)	BCL - Scaled Gasifier and Gas Clean Up Cost $K (2002)
Breault and Morgan [53] [a]	$15,048	$15,801
Dravo Engineering Companies [54] [a]	$14,848	$15,774
Weyerhaeuser, et al., [55] [a]	$21,241	$24,063
Stone & Webster, et al. [56] [a]	$25,067	---
Wan and Malcolm [57] [a]	$18,947 [b] $14,098 [c]	$11,289 [b] $11,109 [c]
Weyerhaeuser [58] [a]	$13,468	$10,224
Wright and Feinberg [59] [a]	$26,048 – BCL design $21,942 – GTI design	$12,318 - quench[d] $26,562 - HGCU [d]
Craig [60]	$13,680	---
AVERAGE	**$18,840**	**$16,392**

(a) From detailed design and cost estimates

(b) Estimated from a 200 dry ton/day plant design.

(c) Estimated from a 1,000 dry ton/day plant design.

(d) Two separate gas clean up configurations were examined for the BCL gasifier. HGCU = hot gas clean up.

Table 17. System Design Information for Gasification References

Reference	Feed Handling and Drying	BCL Gasifier and Gas Clean Up
Breault and Morgan [53]	Rotary dryer	Cyclones, heat exchange & scrubber
Dravo Engineering Companies [54]	Rotary drum dryer	Cyclones, heat exchange & scrubber
Weyerhaeuser, et al. [55]	Steam dryer	Cyclones, heat exchange, tar reformer, & scrubber
Stone & Webster, et al. [56]	Flue gas dryer	---
Wan and Malcolm [57]	Flue gas dryer	Cyclones, heat exchange & scrubber
Weyerhaeuser [58]	Flue gas dryer	Cyclones, heat exchange & scrubber
Wright and Feinberg [59]	Unclear	Quench system – details are not clear Tar reformer system – details are not clear
Craig [60]	Rotary drum dryer	---

For the various pieces of equipment, the design temperature is determined to be the operating temperature plus 50°F (28°C) [63]. The design pressure is the higher of the operating pressure plus 25 psi or the operating pressure times 1.1 [63].

The following costs were estimated based on the Aden, et al. design report: [3]

- cooling tower.
- plant and instrument air.
- steam turbine/generator/condenser package.
- Deaerator.
- alcohol separation equipment (e.g., the distillation columns and molecular sieve unit).

Appendix G contains the design parameters and cost references for the various pieces of equipment in the plant.

4.2. Operating Costs

There are two kinds of operating costs: variable and fixed costs. The following sections discuss the operating costs including the assumptions and values for these costs.

There are many variable operating costs accounted for in this analysis. The variables, information about them, and costs associated with each variable are shown in Table 18.

Previous biomass gasification studies have not looked at fixed operating costs (i.e. salaries, overhead, maintenance, etc) in detail, therefore little data were available. As a result, the fixed operating costs for a biochemical ethanol facility given in Aden, et al., 2002 [3] were used as a starting point to develop fixed costs for this thermochemical design.

The fixed operating costs used in this analysis are shown in Table 19 (labor costs) and Table 20 (other fixed costs). They are shown in 2002 U.S. dollars. The following changes in base salaries and number of employees were made compared to those used in the ethanol plant design in Aden, et al., 2002 [3].

- Plant manager salary raised from $80,000 to $110,000
- Shift supervisor salary raised from $37,000 to $45,000
- Lab technician salary raised from $25,000 to $35,000
- Maintenance technician salary raised from $28,000 to $40,000
- Shift operators salaries raised from $25,000 to $40,000
- Yard employees salaries raised from $20,000 to $25,000 and number reduced from 32 to 12.
- General manager position eliminated
- Clerks and secretaries salaries raised from $20,000 to $25,000 and number reduced from 5 to 3.

The number of yard employees was changed to reflect a different feedstock and feed handling system compared to Aden, et al., 2002 [3]. Handling baled stover requires more hands-on processing when compared to a wood chip feedstock. Based on a 4-shift system, 3 yard employees were estimated to be needed, mostly to run the front end loaders. The general manager position was eliminated because a plant manager would likely be sufficient for this type of facility. Biomass gasification plants are more likely to be operated by larger companies instead of operating like the dry mill ethanol model of farmer co-ops. Finally, the number of clerks and secretaries was reduced from 5 to 3. The estimate of three comes from needing 1 to handle the trucks and scales entering and leaving the facility, 1 to handle accounting matters, and 1 to answer phones, do administrative work, etc.

Since the salaries listed above are not fully loaded (i.e. do not include benefits), a general overhead factor was used. This also covers general plant maintenance, plant security, janitorial services, communications, etc. The 2003 PEP yearbook [71] lists the national average loaded labor rate at $37.66 per hr. Using the salaries in Table 19 above along with the 60% general overhead factor from Aden, et al. [3] gave an average loaded labor rate of $30 per hr. To more closely match the PEP yearbook average, the overhead factor was raised to 95%. The resulting average loaded labor rate was $36 per hr.

Table 18. Variable Operating Costs

Variable	Information and Operating Cost
Tar reformer catalyst	To determine the amount of catalyst inventory, the tar reformer was sized for a gas hourly space velocity (GHSV) of 2,476/hr based on the operation of the tar reformer at NREL's TCPDU where GHSV is measured at standard temperature and pressure [61]. Initial fill then a replacement of 1% per day of the total catalyst volume. Price: $4.67/lb [64]
Alcohol Synthesis Catalyst	Initial fill then replaced every 5 years based on typical catalyst lifetime. Catalyst inventory based on GHSV of 6,000/hr. Price: $5.25/lb [7]
Gasifier bed material	Synthetic olivine and MgO. Delivered to site by truck equipped with self-contained pneumatic unloading equipment. Disposal by landfill. Olivine price: $172.90/ton [65] MgO price: $365/ton [66]
Solids disposal cost	Price: $18/ton [67]
Diesel fuel	Usage: 10 gallon/hr plant wide use Price: $1.00/gallon [68]
Chemicals	Boiler chemicals – Price: $2.80/lb [3] Cooling tower chemicals – Price: $2.00/lb [3] LO-CAT chemicals – Price: $150/tonne of sulfur produced [69]
Waste Water	The waste water is sent off-site for treatment. Price: $2.07/100ft^3 [70]

The updated salaries in Table 19 above were examined against salaries from a free salary estimation tool [72] which uses Bureau of Labor Statistics data and several other sources. Because the biomass analysis does not reflect a specific site in the United States, National Average Salaries for 2003 were used. With such an extensive listing of job titles in the salary estimation tool, a general position such as "clerks and secretaries" could be reflected by multiple job titles. In these instances, care was taken to examine several of the possible job titles that were applicable. A list of the job positions at the production plant and the corresponding job titles in the salary estimation tool [72] is shown in Table 21. Overall, the salaries used in the biomass-tohydrogen production plant design are close to the U.S. national average values given in column 4.

Overall, Aden, et al. [3] lists fixed operating costs totaling $7.54MM in $2000. Using the labor indices, this equates to $7.85MM in $2002. On the other hand, the mixed alcohols design report has fixed operating costs totaling $12.06MM in $2005.

4.3. Value of Higher Alcohol Co-Products

The alcohol synthesis process will create higher molecular weight alcohols. How this co-product is valued will depend upon its end market. There were two extreme cases envisioned. At the high end, these might be sold into the chemical market. This could command a high value for this co-product, upwards to $3.70 to $4.20 per gallon [7]. However, it is unlikely that the market would support more than one or two biomass plants to support these prices. Because of this, the biomass process did not include any detailed separation or clean-up of the separate alcohols. It is envisioned that if this co-product was sold for this purpose, it would be transferred "over the fence" as is and the buyer would take on the costs of separation and clean-up. So, even at the high end, the highest value would be some fraction of the chemical market value.

Table 19. Labor Costs

Position	Salary	Number	Total Cost
Plant manager	$110,000	1	$110,000
Plant engineer	$65,000	1	$65,000
Maintenance supervisor	$60,000	1	$60,000
Lab manager	$50,000	1	$50,000
Shift supervisor	$45,000	5	$225,000
Lab technician	$35,000	2	$70,000
Maintenance technician	$40,000	8	$320,000
Shift operators	$40,000	20	$800,000
Yard employees	$25,000	12	$300,000
Clerks & secretaries	$25,000	3	$75,000
Total salaries (2002 $)			$2,080,000
(2005 $)			$2,270,000

Table 20. Other Fixed Costs

Cost Item	Factor	Cost
General overhead	95% of total salaries	$2,155,000
Maintenance [52]	2% of total project investment	$3,817,000
Insurance & taxes [52]	2% of total project investment	$3,817,000

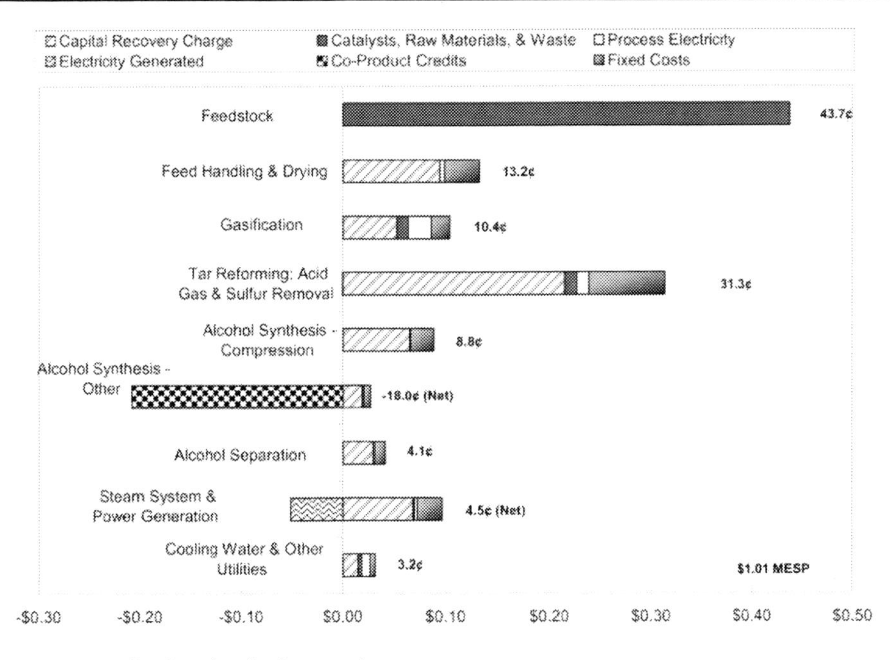

Figure 8. Cost contribution details from each process area

Table 21. Salary Comparison

Job Title in Biomass Plant	Corresponding Job Title in Salary Estimating Tool L72]	Salary Range (17th to 67th percentile)	Average Salary (U.S. national average)	Salary used in Biomass Plant Design (see Table 19)
Plant manager	Plant manager (experience)	$81,042-$220,409	$106,900	$110,000
Plant engineer	Plant engineer	$36,213-$66,542	$58,324	$65,000
Maintenance supervisor	Maintenance crew supervisor	$35,036-$53,099	$45,191	$60,000
	Supervisor maintenance	$34,701-$56,097	$47,046	
	Supervisor maintenance & custodians	$23,087-$45,374	$39,924	
Lab manager	Laboratory manager	$38,697-$70,985	$51,487	$50,000
Shift supervisor	Supervisor production	$32,008-$51 ,745	$43,395	$45,000
Lab technician	Laboratory technician	$25,543-$41 ,005	$34,644	$35,000
Maintenance technician	Maintenance worker	$27,967-$46,754	$39,595	$40,000
Shift operators	Operator control room	$33,983-$61 ,362	$49,243	$40,000
Yard employees	Operator front end loader	$24,805-$39,368	$31,123	$25,000
Clerks & secretaries	Administrative clerk	$19,876-$25,610	$26,157	$25,000
	Secretary	$20,643-$31 ,454	$26,534	
	Clerk general	$1 5,984-$25,61 0	$22,768	

At the low end, the co-product could command a value for a fuel with minimal ASTM standards on its specifications. This would be priced similar to a residual fuel oil. Historically, this is about 80% of gasoline price [73]. Using the ethanol minimum plant gate price as a scaled reference gasoline price (adjusted for ethanol's lower heating value), this translates to \$0.85 per gallon.

For the baseline case, a middle ground was chosen. It is anticipated that the higher alcohols would make an excellent gasoline additive or gasoline replacement in its own right – engine testing and certification would be required. If this is done, then it should command a price similar to that of gasoline. Again using the ethanol minimum plant gate price as a scaled reference gasoline price and adjusting to n-propanol's heating value (the major constituent of the higher alcohol stream), then its value should be \$1.25 per gallon. However, since no special efforts were taken in the process design to clean up this stream to meet anticipated specs, its value is discounted to \$1.15 per gallon.

4.4. Minimum Ethanol Plant Gate Price

Once the capital and operating costs were determined, a minimum ethanol selling price (MESP) was determined using a discounted cash flow rate of return analysis. The methodology used is identical to that used in Aden, et al., (2002) [3]. The MESP is the selling price of ethanol that makes the net present value of the process equal to zero with a 10% discounted cash flow rate of return over a 20 year plant life. The base case economic parameters used in this analysis are given in Table 22. A sensitivity analysis was performed to examine the MESP for different financial scenarios. These are discussed in Section 4.

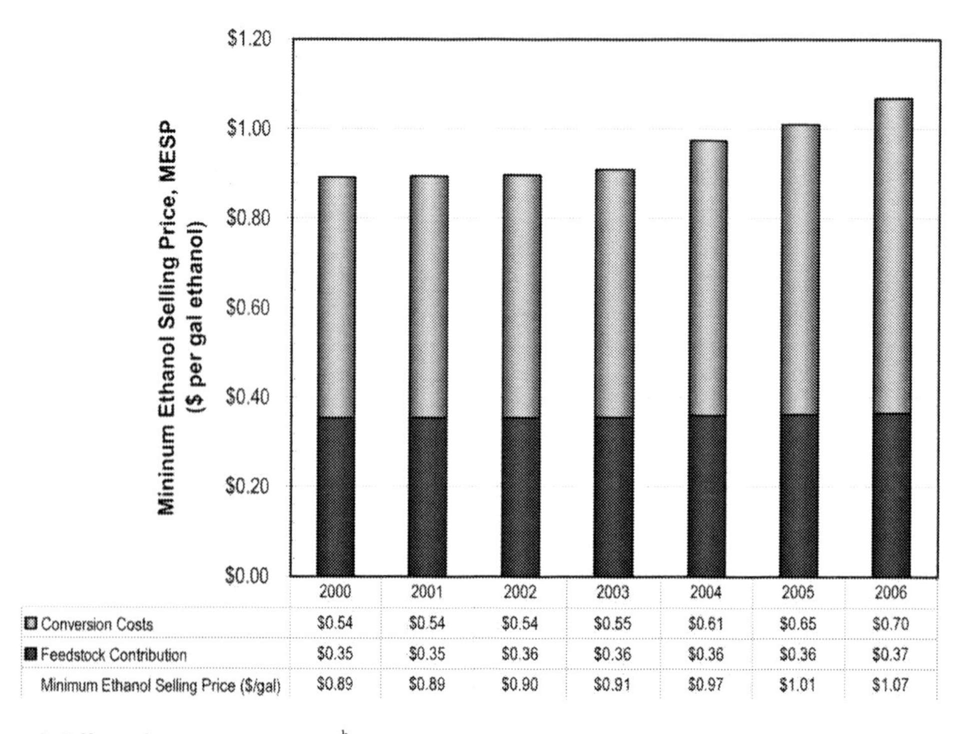

	2000	2001	2002	2003	2004	2005	2006
Conversion Costs	\$0.54	\$0.54	\$0.54	\$0.55	\$0.61	\$0.65	\$0.70
Feedstock Contribution	\$0.35	\$0.35	\$0.36	\$0.36	\$0.36	\$0.36	\$0.37
Minimum Ethanol Selling Price (\$/gal)	\$0.89	\$0.89	\$0.90	\$0.91	\$0.97	\$1.01	\$1.07

Figure 9. Effect of cost year on MESP[h]

Table 22. Economic Parameters

Assumption	Value
Internal rate of return (after-tax)	10%
Debt/equity	0%/100%
Plant life	20 years
General plant depreciation	200% DDB
General plant recovery period	7 years
Steam plant depreciation	150% DDB
Steam plant recovery period	20 years
Construction period 1st 6 months expenditures Next 12 months expenditures Last 12 months expenditures	2.5 years 8% 60% 32%
Start-up time Revenues Variable costs Fixed costs	6 months 50% 75% 100%
Working capital	5% of Total Capital Investment
Land	6% of Total Purchased Equipment Cost (Cost taken as an expense in the 1st construction year)

Note: The depreciation amount was determined using the same method as that documented in Aden, et al. [3] using the IRS Modified Accelerated Cost Recovery System (MACRS).

5. Process Economics, Sensitivity Analyses, and Alternate Scenarios

The cost of ethanol as determined in the previous section was derived using technology that has been developed and demonstrated or is currently being developed as part of the OBP research program. Combined, all process, market, and financial targets in the design represent what must be achieved to obtain the reported $1.01 per gallon. A summary of the breakdown of costs are depicted in Figure 8 and further tabulated in Appendix F.

This cost contribution chart appears to show two different co-product credits: alcohols from the Alcohol Synthesis area and electricity from the Steam System & Power Generation area. However, the process was adjusted so the electricity generated is balanced by the electricity required by all other areas, so there is no net credit for electricity generation.

The cost year chosen for the analysis had a significant effect on the results. As discussed in Section 1.1, capital costs increased significantly after 2003 primarily because of the large increase in steel costs worldwide. Figure 9 depicts how the MESP for this process would change depending on the cost year chosen for the analysis. Notice that between the years 2000 to 2003 the MESP would be much lower, $0.89 to $0.91 per gallon ethanol, instead of the $1.01 determined for 2005. The values for 2006 are tentative, since all factors necessary for the MESP calculation have not yet been published.

The process costs (as indicated by the MESP) are determined from various assumptions on technology (based upon 2012 research targets), markets (such as the value of the higher alcohol co-products), and various financial assumptions (such as required Return on Investment, ROI). When any research target cannot be obtained, or a market or financial assumption does not hold, then the MESP is affected to varying degrees. In addition, uncertainty about equipment design and installation and construction costs will impact the economics. The key is to understand the impact of those types of parameters that are likely to vary, and how they might be controlled to a definable range. Discussed here are process targets that had been identified *a priori* as key ones to understand and achieve. (As can be seen from the sensitivity results, many items examined had much less affect on the MESP than had been thought.) In most cases, values used for the sensitivities are picked from current experimental data, to demonstrate the effect of technology advancement (or lack of) on the economic viability of the process.

The results for the sensitivity analysis discussed in the following sections are depicted in Figure 10; those sensitivities directly impacted by research programs are shown first. Nearly all of these ranges represent variations of a single variable at a time (e.g., ash content while holding the ratio of the non-ash elements constant). There are a couple exceptions to this:

- The feedstock comparison of corn stover to lignin necessitated varying the ultimate elemental analysis, ash content, and moisture content simultaneously.
- The Combined Tar Reformer Conversions incorporated all of the ranges listed for the methane, benzene, and tar simultaneously.

Note that all items in the chart have values associated with them. If a bar is not readily seen, then the MESP effect over the range listed is insignificant.

All analyses are discussed further in the following sections.

5.1. Financial Scenarios

These parameters have the greatest effect on the MESP but R&D has the smallest direct effect on them. In particular, the required ROI for the project could more than double the calculated MESP. Successful R&D and demonstration projects would, at best, ease the ROI requirements of corporations and/or lending institutions and reduce the required MESP toward the baseline case in this report. Also, the baseline of 0% debt financing is not a very realistic scenario, but does represent the conservative endpoint. Many projects of this nature are often financed by some mixture of debt and equity financing. However, the magnitude of this parameter's effect on MESP is quite small in comparison to many of the other financial and market parameters.

A conceptual design like this is normally thought to give accuracy in the capital requirements of -10% to +30%. Using this range for the TPI (Total Project Investment) gives an MESP range of - 6% to +20%.

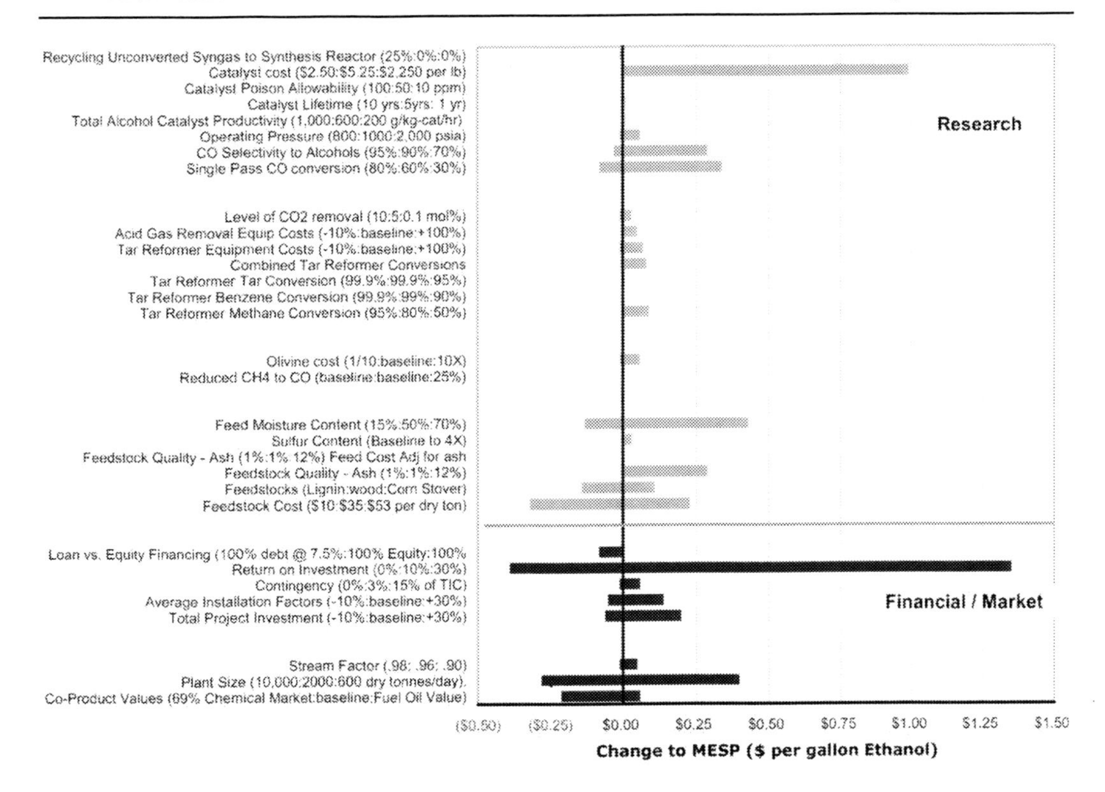

Figure 10. Results of sensitivity analyses

5.2. Feedstocks

Because this process has been designed for utilization of forest resources there may be little control over the feedstock quality coming to the plant[i]. The two most important feedstock quality parameters that can most impact the process economics are moisture and ash content.

The high range of the ash content examined here are more indicative of agricultural residues (from fertilizer) or lignin-rich biochemical process residues; forest resources should have ash contents near that of this baseline case (about 1%). It was originally thought that the cost effects of high ash content could be damped by basing feedstock payments on a dry and ***ash-free*** basis, not just a dry basis. However, Figure 11 shows that this is not the case. Increased ash in the feedstock results in larger ash handling equipment and power requirements, especially in the electrostatic precipitator (ESP) used to remove ash fines from the flue gas. These higher power requirements are met by diverting more syngas to the fuel system to generate electricity. Keeping the feedstock cost constant on a moisture and ash free ("maf") basis decreases the MESP for high-ash feeds by reducing the cost per pound of biomass. However, at a constant mass feedrate to the process, there is inherently less carbon available for conversion to alcohols and therefore smaller revenues. The reduced revenues together with increased capital and operating costs result in an overall increase in MESP despite the lower feedstock cost.

The operating costs due to ash disposal may be reduced by finding an alternate use for the ash. One potential use may be as a soil amendment to replace minerals lost from the soil. The ash collected from gasification in this case should be comparable to the minerals removed from the soil during the plant growth. More study would be needed to determine the best and most economic method for using the ash as a soil amendment.

The biomass feed's moisture content is a problem if it is higher than the baseline 50%. This is not envisioned as being very likely except in the case of processing wet ensiled agricultural residues or energy crops; however, these feedstocks are more envisioned to be processed by biochemical means, not thermochemical means. Drier feedstocks will have lower MESPs because of decreased heat requirements to dry the incoming feedstock directly relate to lower raw syngas diversion to heat and power and higher alcohol yields. This is depicted in Figure 12 and Figure 13. As the moisture content increases, the alcohol yield will decrease because more raw syngas must be diverted for heat. Note that very low moisture contents do not give corresponding increased alcohol yields; this is because flue gas is used for the drying and other operating specifications dictate the amount of raw syngas diverted for heat and power, not the feedstock drying.

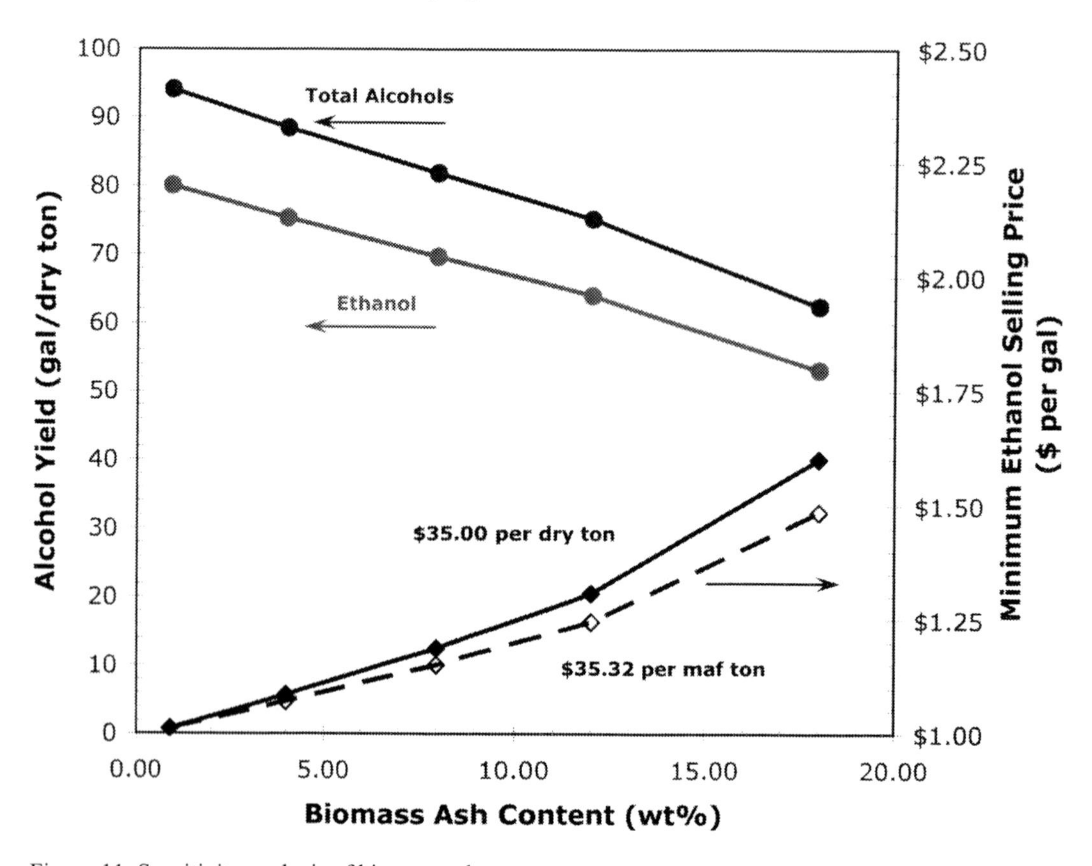

Figure 11. Sensitivity analysis of biomass ash content

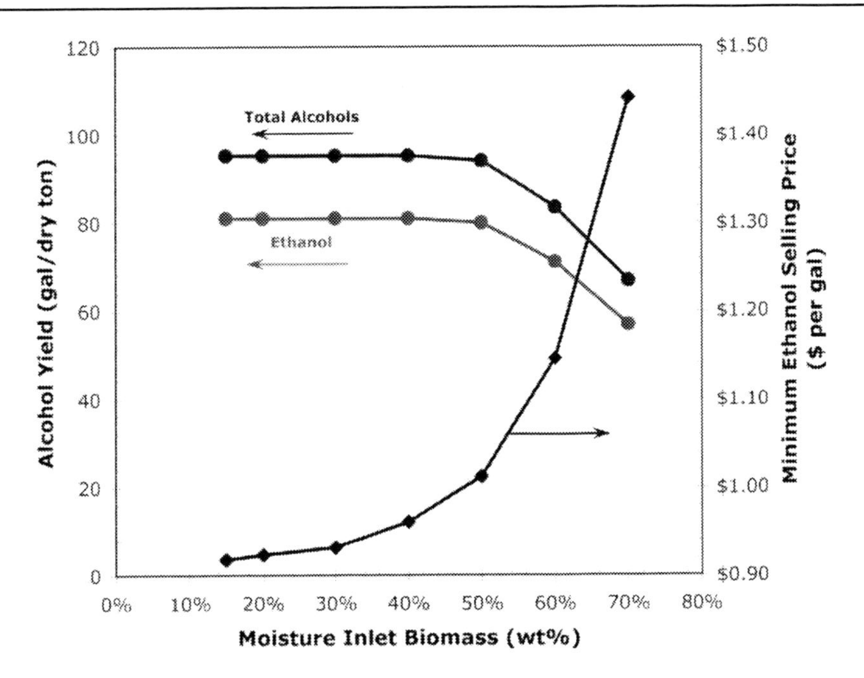

Figure 12. Sensitivity analysis of biomass moisture content

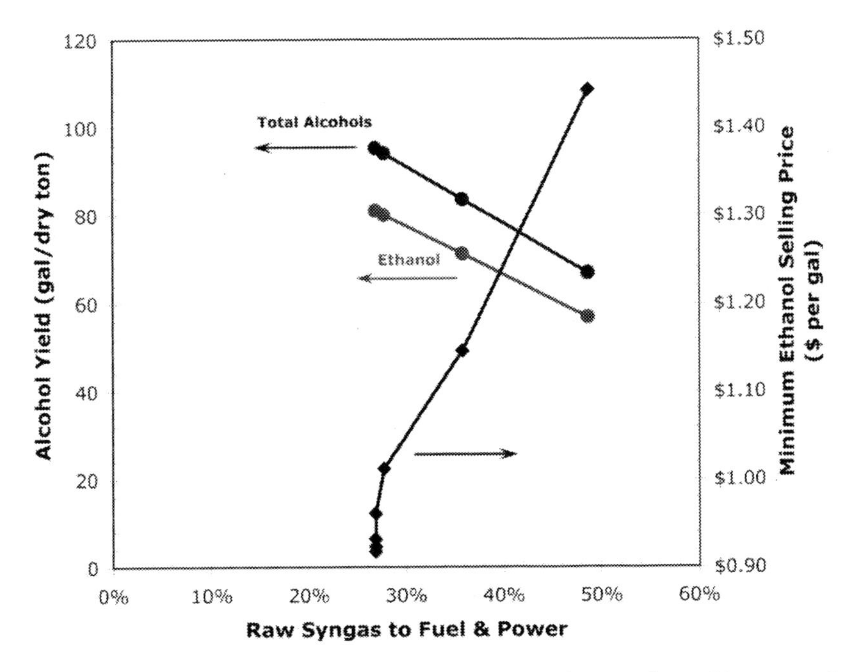

Figure 13. Sensitivity analysis of raw syngas diverted for heat and power due to biomass moisture content

Two combined scenarios were analyzed for two different kinds of feedstocks: corn stover and lignin-rich residues from a biochemical process. The compositions of both are consistent with the Aden et al. design report [3]. Corn stover gives rise to a higher MESP even though its elemental analysis is very similar to wood and its moisture content is very low. The overwhelming effect is due to its higher ash content. Lignin-rich residues have a much lower MESP. Lignin-rich residues also have the virtue of making more electricity than the process needs, so it is exported, even if the raw syngas to the fuel system is minimized while still achieving all other operating specifications. This is a very positive sign that incorporating a thermochemical conversion unit with a biochemical conversion unit and make the heat and power for the entire complex will be cost effective. The feed handling system may have to be different, however, since lignin tends to get very powdery when dried; direct contact with the flue gas for drying would very likely lead to high losses of the feedstock. Drying with indirect contact of the heating medium must be investigated.

5.3. Thermal Conversion

Two gasification scenarios were examined. The first was to explore the impact of increasing the olivine cost which could come about from catalytic modification. Increasing the olivine cost by an order of magnitude could increase the MESP by 6%. This is not significant. The second scenario examined the effect of reducing methane production in the gasifier to reflect a case where the gasifier does not operate within the given correlations. This gave an unexpected result. It was expected that the MESP would decrease when it actually increased a nearly insignificant amount (0.1%). There are two reasons for this. One is an artifact of the way in which the decrease was modeled – more CO was formed and the hydrogen that would have gone to the methane instead went to the char (and was lost for further processing). This required modifications to the operations to keep the H_2:CO ratio to the alcohol synthesis reactor above 1.0 and, in doing so, increased the MESP.

5.4. Clean-Up & Conditioning

These scenarios appeared to have an imperceptible effect on the MESP. However, this is misleading. The scenarios show primarily cost effects due to the material and energy balances. Since the amount of tar is small compared to the amount of CO and H2, these effects are small. In reality, Clean-Up and Conditioning is absolutely required for acceptable performance of gas compressors, waste water treatment, and alcohol synthesis catalysts. Excessive tars in the syngas would significantly impact compressors and waste water treatment with severe consequences to equipment and increased operating costs that are not rigorously modeled here. So, not meeting these targets would give poor performance, leading to greater cost effects than reflected by the sensitivity analysis for this area.

5.5. Fuels Synthesis

These scenarios show the importance of the R&D for the synthesis catalysts. Poor performance could increase MESP by 25% or more. Whether this is due to actual non-target catalyst formulations or due to poor performance in Clean-Up and Conditioning that leads to poor alcohol synthesis catalyst performance, the cost effects are major. The catalyst cost sensitivity range was extremely large, from $2.50/lb to over $2,250/lb. This was done to bracket a variety of potential catalyst systems, not just cobalt moly-sulfide. Exotic metals such as rhodium (Rh) or ruthenium (Ru) can add considerable cost to a catalyst system even at relatively low concentrations. At low catalyst costs, total CO conversion and alcohol selectivity (CO2-free basis) have the largest impact on the overall MESP. The catalyst productivity (g/kg/hr) did not show much impact over the sensitivity range chosen. In reality, all of these catalyst performance indicators are tightly linked. It is unlikely that research could change one without affecting the others.

5.6. Markets

Crediting the co-product higher alcohols with the lower fuel oil value increases the MESP by about 6%; this is still reasonable to meet the qualitative "cost competitive" target. Of even more significance is that selling these higher alcohols for even 69% of their chemical market value will lead to a significant reduction of MESP (about 20%). This shows that the first couple thermochemical conversion plants could get a significant economic advantage in their early life by being able to do this.

6. Conclusions

This analysis shows that biomass-derived ethanol from a thermochemical conversion process has the possibility of being produced in a manner that is "cost competitive with corn-ethanol" by 2012. This thermochemical conversion process would make use of many sub-processes that are currently used commercially (such as acid gas removal) but also requires the successful demonstration of R&D targets being funded by the DOE's OBP.

This analysis has demonstrated that forest resources can be converted to ethanol in a cost competitive manner allowing greater flexibility in converting biomass resources to achieve stated volume targets by 2030.

7. Future Work

Future R&D work to develop and demonstrate reforming and synthesis catalysts is inherent in this study. There many other areas of demonstration and process development also required:

- Demonstrate gasifier performance on other feedstocks (agricultural residues such as corn stover, energy crops such as switchgrass, and lignin-rich residues that would be available from a co-located biochemical conversion process). Of particular importance for the lignin-rich residues is the impact on process performance of trace amounts of chemicals used in the biochemical processing that might negatively impact the thermochemical conversion process.
- Compare the relative merits of direct oxygen blown gasifiers to the indirect steam gasifier upon which this study is based.

- Examine the trade-offs of the greater use of water cooling (greater water losses in the cooling tower) vs. air cooling (greater power usage) vs. organic Rankine cycle for cooling and power production.
- Better understand the trade offs between operating conditions in the alcohol synthesis reactor to operating conditions (pressure, temperature, extent of reaction, extent of methanol recycle). A "tuned" kinetics based model would be required for this.
- Explore alternate synthesis reactor configurations (slurry phase vs. fixed bed).
- Understand trade offs between an energy neutral, alcohol production facility to one that could also supply heat and electricity to a co-located biochemical conversion facility.
- Further explore the potential benefits of integrating biochemical and thermochemical technologies.
- Examine potential for decreased heat integration complexity and increased overall energy efficiency.
- Better understand the kinetics of catalytic tar reforming and deactivation, and the necessary regeneration kinetics to achieve a sustainable tar reforming process.
- Examine the emissions profile from the plant and explore alternate emissions control equipment.

REFERENCES

[1] 2006 State of the Union address, January 31, 2006. http://www.whitehouse.gov/stateoftheunion/2006/

[2] http://www.whitehouse.gov/news/releases/2006/01/20060131-6.html

[3] Aden, A; Ruth, M; Ibsen, K; Jechura, J; Neeves, K; Sheehan, J & Wallace, B; Montague, L.; Slayton, A.; Lukas, J. *Lignocellulosic Biomass to Ethanol Process Design and Economics Utilizing Co-Current Dilute Acid Prehydrolysis and Enzymatic Hydrolysis for Corn Stover.* NREL Report No. TP-510-32438. June 2002. http://www.nrel.gov/docs/fy02osti/32438.pdf

[4] CEH Marketing Research Report - Ethyl Alcohol, *Chemical Economics Handbook — SRI International*, pp 50-51. 2002

[5] Wooley, R; Ruth, M.; Sheehan, J & Ibsen, K. *Lignocellulosic Biomass to Ethanol Process Design and Economics Utilizing Co-Current Dilute Acid Prehydrolysis and Enzymatic Hydrolysis Current and Futuristic Scenarios*, NREL report TP-580-26157, July 1999. http://devafdc.nrel.gov/pdfs/3957.pdf

[6] Perlack, RD; Wright, LL; Turhollow, AF; Graham, RL; Stokes, BJ & Erbach, DC. *Biomass as Feedstock for a Bioenergy and Bioproducts Industry: the Technical Feasibility of a Billion-Ton Annual Supply.* A joint U.S. Department of Energy and U.S. Department of Agriculture report. DOE/GO-102995-2135 & ORNL/TM-2005/66. April 2005. http://www1.eere.energy billionton vision report2.pdf

[7] Aden, A; Spath, P & Atherton, B. *The Potential of Thermochemical Ethanol Via Mixed Alcohols Production.* NREL Milestone Report. September 30, 2005. http://devafdc.nrel.gov/bcfcdoc/9432.pdf

[8] Bain, RL; *The Role of Thermochemical Processing in an Integrated Sugars/Thermochemical Biorefinery: Conceptual Process Technoeconomic Analysis.* NREL Technical Memo, February 9, 2005. http://devafdc.nrel.gov/bcfcdoc/8967.pdf

[9] Spath, P; Aden, A; Eggeman, T; Ringer, M; Wallace, B & Jechura, J. *Biomass to Hydrogen Production Detailed Design and Economics Utilizing the Battelle Columbus Laboratory Indirectly-Heated Gasifier.* NREL Report No. Number NREL/TP-510-37408. May 30, 2005. http://www.nrel.gov/docs/fy05osti/37408.pdf

[10] http://engineering.dartmouth.edu/rbaef/ Final report under development.

[11] Larson, ED; Jin, H & Celika, FE. "Large-Scale Gasification-Based Co-Production of Fuels and Electricity from Switchgrass." Draft manuscript for *Biomass & Bioenergy.* March 2006.

[12] Larson, ED; McDonald, GW; Yang, W; Frederick, WJ; Iisa, K; Kreutz, TG; Malcom, EW; & Brown, CA; "A Cost-Benefit Assessment of Black Liquor Gasifier/Combined Cycle Technology Integrated into a Kraft Pulp Mill," Tappi Journal, 83(6): 57-, June 2000.

[13] Bain, RL; Craig, KR & Overend, RP; (2000). "Gasification for Heat and Power, Methanol, and Hydrogen," Chapter 9.2 of *Industrial Uses of Biomass Energy*, ed. F. RosilloCalle, et al., Taylor and France, London, UK, ISBN-0-7484-0884-3.

[14] Wyman, CE; Bain, RL; Hinman, ND; & Stevens, DJ; (1993) "Ethanol and Methanol from Cellulosic Materials, "Chapter 21 of *Renewable Energy: Sources for Fuels and Electricity*, ed T.B. Johansson, et al, Island Press.

[15] Chem Systems "*Assessment of Costs and Benefits of Flexible and Alternative Fuel Use in the U.S. Transportation Sector*", by Chem Systems, Tarrytown, NY, for the U.S. Department of Energy, Washington, D.C., Report No. DOE/PE--0093, 46 pp, November 1989.

[16] Feldmann, HF; Paisley, MA; Appelbaum, HR & Taylor, DR; "Conversion of Forest Residues to a Methane-Rich Gas in a High- Throughput Gasifier", prepared by Battelle Columbus Division, Columbus, Ohio, for the Pacific Northwest Laboratory, Richland, WA, PNL Report no. PNL-6570, May 1988.

[17] Wan, EI; & Fraser, MD; "Economic Assessment of Advanced Biomass Gasification Systems", by Science Applications International Corporation, McLean, VA, IGT Biomass Conference, New Orleans, LA, Feb 1989.

[18] Wyman, CE; Bain, RL; Hinman, ND & Stevens, DJ; (1993) "Ethanol and Methanol from Cellulosic Materials, "Chapter 21 of *Renewable Energy: Sources for Fuels and Electricity*, ed T.B. Johansson, et al, Island Press.

[19] Hamelinck, CN & Faaij, APC; (2001). "*Future Prospects for Production of Methanol and Hydrogen from Biomass*," Utrecht University, Utrecht, the Netherlands, Report No. NWS-E2001-49, ISBN-90-73958-84-9

[20] Williams, RH; Larson, ED; Katofsky, RE & Chen, J; (1995). "*Methanol and Hydrogen from Biomass for Transportation, with Comparisons to Methanol and Hydrogen from Natural Gas and Coal*," PU/CEES Report No. 292, Princeton University, Princeton, NJ, July.

[21] Nexant Inc. *Equipment Design and Cost Estimation for Small Modular Biomass Systems, Synthesis Gas Cleanup, and Oxygen Separation Equipment, Task 1. Cost Estimates of Small Modular Systems.* Subcontract Report NREL/SR-510-39943. May 2006. http://www.nrel.gov/docs/fy06osti/39943.pdf

[22] Nexant Inc. *Equipment Design and Cost Estimation for Small Modular Biomass Systems, Synthesis Gas Cleanup, and Oxygen Separation Equipment ,Task 2. Gas Cleanup Design and Cost Estimates – Wood Feedstock.* Subcontract Report NREL/SR-510-39945. May 2006. http://www.nrel.gov/docs/fy06osti/39945.pdf

[23] Nexant Inc. *Equipment Design and Cost Estimation for Small Modular Biomass Systems, Synthesis Gas Cleanup, and Oxygen Separation Equipment, Task 2.3: Sulfur Primer.* Subcontract Report NREL/SR-510-39946 May 2006. http://www.nrel.gov/docs/fy06osti/39946.pdf

[24] Nexant Inc. *Equipment Design and Cost Estimation for Small Modular Biomass Systems, Synthesis Gas Cleanup, and Oxygen Separation Equipment, Task 9: Mixed Alcohols From Syngas — State of Technology.* Subcontract Report NREL/SR-510-39947 May 2006. http://www.nrel.gov/docs/fy06osti/39947.pdf

[25] Garrett, DE; *Chemical Engineering Economics*, Van Nostrand Reinhold, New York, 1989.

[26] Peters, MS; Timmerhaus, KD; *Plant Design and Economics for Chemical Engineers*, 5th Edition, McGraw-Hill, Inc., New York. 2003.

[27] Perry, RH; Green, DW; Maloney, J.O. *Perry's Chemical Engineers' Handbook*, 7th ed., McGraw-Hill. 1997

[28] October 2006 values used from http://www.che.com/inc/CEeconomicIndicators.php.

[29] Craig, K.R.; Mann, M.K. *Cost and Performance Analysis of Biomass-Based Integrated Gasification Combined-Cycle (BIGCC) Power Systems*, NREL Report No. NREL/TP-430- 21657. October 1996. http://www.nrel.gov/docs/legosti/fy97/21657.pdf

[30] Foust, T; et al., *A National Laboratory Market and Technology Assessment of the 30x30 Scenario*. Draft in progress, 2007, DOE Report.

[31] Mann, MK; Spath, P.L. *Life Cycle Assessment of a Biomass Gasification Combined-Cycle Power System.* National Renewable Energy Laboratory, Golden, CO, TP-430-23076. 1997. http://www.nrel.gov/docs/legosti/fy98/23076.pdf

[32] Phillips, S; Carpenter, Dayton, D; Feik, D; French, C; Ratcliff, R; Hansen, M; Deutch, R; & Michener, SB. (2004). *Preliminary Report on the Performance of Full Stream Tar Reformer.* Internal NREL Milestone report.

[33] Nexant Inc. *Equipment Design and Cost Estimation for Small Modular Biomass Systems, Synthesis Gas Cleanup, and Oxygen Separation Equipment ,Task 2: Gas Cleanup Design and Cost Estimates – Wood Feedstock.* Subcontract Report NREL/SR-510-39945. May 2006. http://www.nrel.gov/docs/fy06osti/39945.pdf

[34] Gas Processors Suppliers Association. *Engineering Data Book*, FPS Version, 12th ed., Tulsa, OK. 2004

[35] A Claus unit for reducing sulfur was briefly considered as an alternative to LO-CAT, however the size of the stream and H2S concentration made this a fairly impractical choice.

[36] Nexant, Inc. *Equipment Design and Cost Estimation for Small Modular Biomass Systems, Synthesis Gas Cleanup, and Oxygen Separation Equipment. Task 9: Mixed Alcohols from Syngas – State of Technology.* NREL/SR-510-39947. Performed by Nexant Inc., San Fransisco, CA. Golden, CO: National Renewable Energy Laboratory, May 2006. http://www.nrel.gov/docs/fy06osti/39947.pdf

[37] Herman, RG; (1991). "Chapter 7 – Classical and Non-classical Routes for Alcohol Synthesis." New Trends in CO Activation, L. Guczi, ed., Elsevier, New York, pp 265-349.
[38] Consult Appendix J for specific literature experiment.
[39] Personal communication between Yves Parent at NREL and CRITERION (Houston, TX) on Friday, April 13, 2006.
[40] Email correspondence between David Dayton, NREL, and Mark Jones, Dow Chemical. January 19, 2007.
[41] Forzatti, P; Tronconi, E & Pasquon, I; (1991). "Higher Alcohol Synthesis." *Catalysis Reviews – Science and Engineering,* 33(1-2), pp109-168.
[42] Hamelinck, C; et al. *Production of FT Transportation Fuels from Biomass; Technical Options, Process Analysis, and Optimisation, and Development Potential.* Utrecht University, Copernicus Institute. March 2003. ISBN 90-393-3342-4.
[43] Quarderer, GJ; "Mixed Alcohols from Synthesis Gas." *Proceedings from the 78th Spring National AIChE Meeting.* April 6-10, 1986, New Orleans, LA.
[44] Nirula, S; "Dow / Union Carbide Process for Mixed Alcohols from Syngas" SRI International. March 1986. PEP Review No 85-1-4. Menlo Park, CA.
[45] Park, T; et al.; "Kinetic Analysis of Mixed Alcohol Synthesis from Syngas over K/MoS2 Catalyst." *Ind. Eng. Chem. Res.* Vol 36, 1997. pp 5246 – 5257.
[46] Gunturu, A; et al. "A Kinetic Model for the Synthesis of High-Molecular-Weight Alcohols over a Sulfided Co-K-Mo/C Catalyst." *Ind. Eng. Chem. Res.* Vol. 37, 1998. pp. 2107 – 2115.
[47] Smith, K; et al. "Kinetic Modeling of Higher Alcohol Synthesis over Alkali-Promoted Cu/ZnO and MoS2 Catalysts." *Chemical Engineering Science*, Vol. 45. 1990. pp. 2639 – 2646.
[48] Gas Processors Suppliers Association. *Engineering Data Book*, FPS Version, 12th ed., Tulsa, OK. 2004
[49] Weast, RC; ed. *CRC Handbook of Chemistry and Physics*, 62nd ed., CRC Press, Boca Raton, FL. 1981
[50] Aden, et .al. "Lignocellulosic Biomass to Ethanol Process Design and Economics Utilizing Co-Current Dilute Acid Prehydrolysis and Enzymatic Hydrolysis for Corn Stover", NREL/TP510-32438, June 2002.
[51] Keeney, D; et al. "Water Use by Ethanol Plants Potential Challenges". Institute for Agriculture and Trade Policy, Minneapolis, MN. October 2006. http://www.iatp.org/iatp/publications.cfm?accountID=258&refID=89449
[52] Peters, MS; Timmerhaus, KD; *Plant Design and Economics for Chemical Engineers*, 5th Edition, McGraw-Hill, Inc., New York. 2003.
[53] Breault, R & Morgan, D; *Design and Economics of Electricity Production Form An Indirectly- heated Biomass Gasifier.* Report TR4533-049-92. Columbus, Ohio: Battelle Columbus Laboratory. 1992.
[54] Dravo Engineering Companies. Gasification Capital Cost Estimation. Obtained from Mark Paisley, August, 1994. Battelle Columbus Laboratory. 1987.
[55] Weyerhaeuser, Nexant, and Stone & Webster. *Biomass Gasification Combined Cycle.* Weyerhaeuser Company, Tacoma, WA . DOE DE-FC36-96GO10173. 2000.

[56] Stone & Webster; Weyerhaeuser; Amoco; and Carolina Power & Light. *New Bern Biomass to Energy Project Phase 1 Feasibility Study*. Response to NREL Contract No. LOI No. RCA-3- 13326. NREL Report No. TP-421-7942. June 1995.

[57] Wan, EI & Malcolm, DF; "Economic Assessment of Advanced Biomass Gasification Systems," in *Energy from Biomass and Wastes XIII*, Donald L. Klass, ed. Chicago: Institute of Gas Technology, pp.791-827. 1990.

[58] Weyerhaeuser. Gasification Capital Cost Estimation. Obtained from Mark Paisley, August, 1994. Battelle Columbus Laboratory. 1992.

[59] Wright, J & Feinberg, DA; *Comparison of the Production of Methanol and Ethanol from Biomass*. For the International Energy Agency. Contract no. 23218-1-9201/01-SQ. 1993.

[60] Craig, K; Electric Power Generation Cost - Version 1.11 spreadsheet from Kevin Craig. July 6, 1994.

[61] Fogler, HS; *Elements of Chemical Reaction Engineering*. Second Edition. Prentice Hall. Englewood Cliffs, New Jersey. 1992.

[62] Kohl, AL & Nielsen, RB; *Gas Purification*, 5th Edition. Gulf Publishing Company. 1997

[63] Walas, SM; *Chemical Process Equipment Selection and Design*. Butterworth-Heinemann. 1988.

[64] Leiby, SM. (1994) *Operations for Refinery Hydrogen*. SRI Report No. 212. Menlo Park, CA.

[65] Jeakel, D. (2004). Price quote from AGSCO for super sacks or bulk.

[66] *Chemical Marketing Reporter*. (2004). August 23-30 issues.

[67] Chem Systems. (1994). Biomass *to Ethanol Process Evaluation*. Prepared for NREL. Tarrytown, New York.

[68] Energy Information Agency (EIA). 2003. http://www.eia.doe.gov/pub/oil_gas/ petroleum/data_publications/petroleum_marketing_monthly /current/pdf/pmmtab 1 6.pdf.

[69] Graubard, D. Personal correspondence. Gas Technology Products LLC. Schaumburg, IL. 2004

[70] East Bay municipal utility district. 2004. http://www.ebmud.com/wastewater/ industrial_&_commercial_permits_&_fees/wastewater_rates /default.htm#non-residential%20rates

[71] SRI International. (2003). *PEP Yearbook International*. Volume 1E. United States. Menlo Park, CA.

[72] Baker, Thomsen Associates Insurance Services Inc. (BTA), *Salary Expert ePro©, www.salaryexpert.com, 2004.*

[73] Based on historical wholesale prices found on the Energy Information Agency web site, http://www.eia.doe.gov/oil gas/petroleum/info glance/petroleum.html

End Notes

[a] The curve marked "Ethyl Alcohol" is for 190 proof, USP, tax-free, in tanks, delivered to the East Coast. That marked "Specially Denatured Alcohol" is for SDA 29, in tanks, delivered to the East

Coast, and denatured with ethyl acetate. That marked "Fuel Alcohol" is for 200 proof, fob works, bulk, and denatured with gasoline

[a*] Calcined magnesium silicate, primarily Enstatite ($MgSiO_3$), Forsterite (Mg_2SiO_3), and Hematite (Fe_2O_3). This is used as a sand for various applications. A small amount of magnesium oxide (MgO) is added to the fresh olivine to prevent the formation of glass-like bed agglomerations that would result from biomass potassium interacting with the silicate compounds.

[c] Calculated using the Aspen Plus Boie correlation.

[d] Higher Heating Value

[e] Lower Heating Value

[f] Converts to N_2 and H_2.

[g] Based on assumed catalyst density of 64 lb/ft^3, 600 g/kg-catalyst/hr = 615 g/L-catalyst/hr.

[h] Note that the relative splits between feedstock and conversion costs have been scaled to attribute some of the costs to the mixed alcohol co-products. So, the feedstock contribution appears to be different than what is depicted in the cost contribution chart for the different areas.

[i] At least less so than using agricultural residues or energy crops that can be bred for specific properties in these lignocellulosic materials.

APPENDIX A. LIST OF ACRONYMS

ASME	American Society of Mechanical Engineers
BCL	Battelle Columbus Laboratory
BFW	Boiler Feed Water
bpd	Barrels per Day
BTU	British Thermal Unit
CFM	Cubic Feet per Minute
CH4	Methane
CIP	Clean-in-place
CO	Carbon Monoxide
Co	Cobalt
CO_2	Carbon Dioxide
DCFROR	Discounted Cash Flow Rate of Return
DOE	US. Department of Energy
EIA	Energy Information Administration
EtOH	Ethanol
FT	Fischer-Tropsch
FY	Fiscal Year
GHSV	Gas Hourly Space Velocity
GJ	GigaJoule
gpm	Gallons per minute
H2	Hydrogen
HAS	Higher Alcohol Synthesis
HHV	Higher Heating Value
IFP	Institut Francais du Petrole
IRR kWh	Internal Rate of Return Kilowatt-hour

LHV	Lower Heating Value
MA	Mixed Alcohols
MASP	Minimum alcohols selling price
MeOH	Methanol
MESP	Minimum ethanol selling price
MoS_2	Molybdenum disulfide
MTBE	Methyl-tertiary-butyl Ether
MW	Megawatts
NREL	National Renewable Energy Laboratory
NRTL	Non-Random Two Liquid activity coefficient method
OBP	Office of the Biomass Program
PFD	Process flow diagram
PEFI	Power Energy Fuels, Inc.
PNNL	Pacific Northwest National Laboratory
PPMV	Parts per million by volume
psia	Pounds per square inch (absolute)
RKS-BM	Redlich-Kwong-Soave equation of state with Boston-Mathius modifications
SEHT	Snamprogetti, Enichem and Haldor Topsoe
SMR	Steam Methane Reformer
TC EtOH	Thermochemical Ethanol
tpd	Short Tons per Day
TPI	Total Project Investment
UCC	Union Carbide Corp.
WGS	Water Gas Shift
WRI	Western Research Institute
WWT	Wastewater Treatment

Appendix B. OBP Thermochemical Platform Research Targets

Background/Introduction

Thermochemical conversion technology options include both gasification and pyrolysis. Thermochemical conversion is envisioned to be important in enabling lignocellulosic biorefineries and to maximize biomass resource utilization for the production of biofuels. Moving forward, the role of thermochemical conversion is to provide a technology option for improving the economic viability of the developing bioenergy industry by converting the fraction of the biomass resources that are not amenable to biochemical conversion technologies into liquid transportation fuels. The thermochemical route to ethanol is synergistic with the biochemical conversion route. A thermochemical process can more easily convert low- carbohydrate or "non-fermentable" biomass materials such as forest and wood residues to alcohol fuels, which adds technology robustness to efforts to achieve the 30 x 30 goal (Foust, et al., 2006). This Appendix describes the R&D needed to achieve the market target production price in 2012 for a stand-alone biomass gasification/mixed alcohol process. Future advanced technology scenarios rely on considerable biofuel yield enhancements achieved by combining biochemical and thermochemical conversion technologies into an integrated biorefinery that implements mixed alcohol production from gasification of lignin-rich bioconversion residues to maximize the liquid fuel yield per delivered ton of biomass.

Biomass gasification can convert a heterogeneous supply of biomass feedstock into a consistent gaseous intermediate that can then be reliably converted to liquid fuels. The biomass gasification product gas ("synthesis gas" or simply "syngas") has a low to medium energy content (depending on the gasifying agent) and consists mainly of CO, H_2, CO_2, H_2O, N_2, and hydrocarbons. Minor components of the syngas include tars, sulfur and nitrogen oxides, alkali metals, and particulates. These minor components of the syngas potentially threaten the successful application of downstream syngas conversion steps.

Commercially available and near-commercial syngas conversion processes were evaluated on technological, environmental, and economic bases (Spath and Dayton, 2003). This design report provides the basis for identifying promising, cost-effective fuel synthesis technologies that maximize the impact of biomass gasification for transforming biomass resources into clean, affordable, and domestically produced biofuels. For the purpose of this report the pre- commercial mixed alcohols synthesis process implementing an alkali promoted MoS2 catalyst, a variant of Fischer-Tropsch synthesis, was selected as the conversion technology of choice because high yields of ethanol are possible with targeted R&D technology advancements. The MoS2 catalyst is also tolerant of low levels of sulfur gases that are common catalyst poisons. The proposed mixed alcohol process does not produce ethanol with 100% selectivity. Production of higher normal alcohols (e.g., n-propanol, n-butanol, and n-pentanol) is unavoidable. Fortunately, these by-product higher alcohols have value as commodity chemicals, fuel additives, or potentially fuels in their own right.

The schedule for meeting specific research goals for improved tar reforming and mixed alcohol synthesis catalyst performance was accelerated by the President's Advanced Energy Initiative to achieve cost-competitive cellulosic ethanol by 2012. This design report provides a rigorous engineering analysis to provide a baseline technology scenario for doing this. The conceptual process design and ethanol production cost estimate quantify the benefits of meeting the R&D goals for tar reforming and improved mixed alcohol catalyst performance helps establish technical R&D targets that need to be overcome by a concerted and directed core research effort.

Process Description

Figure 1 shows a block process flow diagram of the cost-competitive target process and the major technical barriers that need to be addressed to accomplish this target case. The feedstock interface addresses the main biomass fuel properties that impact the long-term technical and economic success of a thermochemical conversion process: moisture content, fixed carbon and volatiles content, impurity (sulfur, nitrogen, chlorine) concentrations, and ash content. High moisture and ash contents reduce the usable fraction of delivered biomass fuels proportionally. Therefore, maximum system efficiencies should be possible with dry, low ash biomass fuels.

Biomass gasification is a complex thermochemical process that begins with the thermal decomposition of a lignocellulosic fuel followed by partial oxidation of the fuel with a gasifying agent, usually air, oxygen, or steam to yield a raw syngas. The raw gas composition and quality are dependent on a wide range of factors including feedstock composition, type of gasification reactor, gasification agents, stoichiometry, temperature, pressure, and the presence or lack of catalysts.

Gas cleanup is a general term for removing the unwanted impurities from biomass gasification product gas and generally involves an integrated, multi-step approach that depends on the end use of the product gas. This entails removing or eliminating tars, acid gas removal, ammonia scrubbing, alkali metal capture, and particulate removal. Gas conditioning refers to final modifications to the gas composition that makes it suitable for use in a fuel synthesis process. Typical gas conditioning steps include sulfur polishing to remove trace levels of remaining H_2S and water-gas shift to adjust the final H_2:CO ratio for optimized fuel synthesis.

Comprehensive cleanup and conditioning of the raw biomass gasification product gas yields a "clean" syngas comprised of essentially CO and H_2, in a given ratio that can converted to a mixed alcohol product. Separation of ethanol from this product yields a methanol-rich stream that can be recycled with unconverted syngas to improve process yield. The higher alcohol-rich stream yields by-product chemical alcohols. The fuel synthesis step is exothermic so heat recovery is essential to maximize process efficiency.

R&D Needs To Achieve the 2012 Technical Target for Thermochemical Ethanol

Essential R&D activities from 2007 through 2011 to overcome identified technical barrier areas to meet the established 2012 technical target for thermochemical ethanol production are outlined in Table 1. The rigorous engineering analysis of the thermochemical ethanol process conducted in this study will help to validate the feasibility of these technical targets and provide focus for the technical barriers that provide the largest economic benefit. These R&D activities include fundamental kinetic measurements, micro-activity catalyst testing, bench-scale thermochemical conversion studies, pilot-scale validation of tar reforming catalyst performance, mixed alcohol catalyst development, and pilot-scale demonstration of integrated biomass gasification mixed alcohol synthesis. Process data collected in the integrated pilot-scale testing will provide the basis for process optimization and cost estimates that will guide deployment of the technology.

Feedstock Interface

Feedstock handling, processing, and feeding specifically related to the thermochemical conversion process will need to be addressed. Because the 30 x 30 scenario envisions mixed alcohol conversion for low-grade or "non-fermentable" feedstocks, refinements in dry biomass feeder systems for use with gasification will be required to meet cost targets. These refinements should reduce upfront feed processing requirements to yield biomass feedstocks at $35 per ton delivered to the thermochemical process. Additional challenges will be associated with feeding the delivered biomass into developing pressurized biomass gasification systems. In all cases, demonstrating biomass feed systems beyond the pilot-scale will be necessary but this is not a significant component of the proposed research portfolio.

Fundamental Gasification Studies R&D Needs

The thermochemical mixed alcohol synthesis conversion route is envisioned initially for forest thinnings and other predominately woody feedstocks and residues. Hence, gasification studies will need to be performed to determine how feedstock composition affects syngas composition and quality and syngas efficiency. The gasifier technology chosen for the basis of this analysis is the Battelle Columbus Laboratory indirectly heated gasifier. Other gasifier technologies are under development that could prove more promising. These technologies will need to be tracked to ascertain their applicability to the mixed alcohol synthesis process.

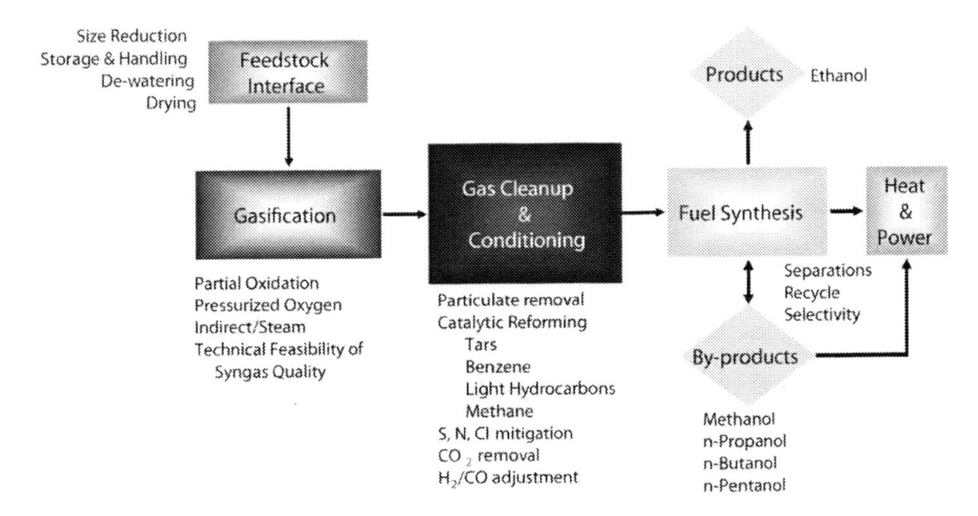

Figure 1. Process flow diagram with research barriers for cost-competitive thermochemical ethanol production

Tar Cleanup and Conditioning R&D Needs

Previous techno-economic analyses (Aden and Spath, 2005) have shown that achieving the research goals for cleanup and conditioning of biomass-derived syngas to remove chemical contaminants such as tar, ammonia, chlorine, sulfur, alkali metals, and particulates has the greatest impact on reducing the cost of mixed alcohol synthesis. To date, gas cleanup and conditioning technologies and systems are unproven in integrated biorefinery applications. The goal of this research is to eliminate the tar removal and disposal via water quench, which is problematic both from efficiency and waste disposal perspectives, and develop a consolidated tar and light hydrocarbon reforming case.

Table 1. Thermochemical Ethanol (Gasification/Mixed Alcohols) R&D Targets to meet the 2012 Cost-Competitive Thermochemical Ethanol Cost Target

R&D Area	Current	2007	2008	2009	2010	2011	2012
Feedstock Interface	\$30/dry ton wood chips 50% moisture dried to 12% - 2000 tpd plant					\$30/dry ton biorefinery residues based on \$45/dry ton corn stover. 50% moisture dried to 12% - 2000 tpd plant	
Thermo-chemical Conversion - Gasification	Wood chips (model) - Indirect (atm) gasification – 78% syngas efficiency: H_2/CO = 1.0-1.5 $CH_4 \leq 15$vol% Tars ≤ 30 g/Nm^3 benzene $\leq$ 1vol% H_2S = 50-600 ppm NH_3 and HCl to be determined	Biorefinery residues - Indirect (atm) gasification : corn stover; switchgrass; wheat straw; lignin - 78% syngas efficiency: H_2/CO = 1.0-1.5 $CH_4 \leq 15$vol% Tars $\leq$ 30 g/Nm^3; benzene $\leq$ 1vol% H_2S = 50-600 ppm NH_3 and HCl to be determined	Demonstrate biomass gasification for \$6.88/MMBtu syngas cost based on 2007	Indirect (atm) gasification – 78% syngas efficiency: H_2/CO = 1.0-1.5 $CH_4 \leq 8$vol% Tars $\leq$10 g/Nm^3; benzene < 0.1vol%; $H_2S \leq$ 20 ppm; NH_3 and HCl to be determined	Demonstrate biomass gasification for \$5.25/ MMBtu syngas cost	Indirect (atm) catalytic gasification – 78% syngas efficiency: H_2/CO = 1.0 $CH_4 \leq$ 5vol% Tars $\leq$ 1 g/Nm^3; benzene $\leq$ 0.04 vol%; $H_2S \leq$ 20 ppm; NH_3 and HCl to be determined	

Table 1. (Continued)

R&D Area	Current	2007	2008	2009	2010	2011	2012
Cleanup and Conditioning	Cyclone particulate removal $H_2S \geq$ 50 ppm (based on feedstock) with no S removal Tar Reformer Efficiency $CH_4 \geq$ 20%	Sorbent injection to maintain H2S levels $\leq$ 50 ppm for syngas from biomass to reduce sulfur deactivation of tar reforming catalysts.	Tar Reformer Efficiency $CH_4 \geq 50\%$ Benzene $\geq 90\%$ heavy tars $\geq 97\%$ (79% CH_4 conversion in separate SMR)	Improve tar reforming catalyst performance - Regen/TOS ratio $\leq$ 600	Tar Reformer Efficiency $CH_4 \geq 80\%$ Benzene $\geq 99\%$ heavy tars $\geq 99.9\%$ Eliminate SMR; highest activity re-gained by regenerating deactivated	Improve tar reforming catalyst performance - Regen/TOS ratio $\leq$ 250	Integrated operations for syngas cleanup and conditioning target composition for fuel; synthesis: $CH_4 \leq 3$vol% Benzene ≤ 10 ppm Heavy tars ≤ 0.1 g/Nm^3 $H_2S \leq 1$ ppm $NH_3 \leq 10$ ppm HCl $\leq$ 10 ppb
Catalytic Fuels Synthesis (Mixed Alcohols)	$H_2/CO = 1.2$ Pressure $\leq$ 2000 psia Productivity = 100-400 gMA/ kg(cat)/hr EtOH Selectivity $\geq 70\%$ (CO_2-free)	$H_2/CO \leq 1.2$ Pressure $\leq$ 2000 psia Productivity $\geq$ 150 gMA/ kg(cat)/hr EtOH Selectivity $\geq 70\%$ (CO_2-free)	Demonstrate 500 hours catalyst lifetime at 2007 performance with bottled syngas for mixed alcohol catalyst cost of $\leq$\$0.50/gal EtOH	$H_2/CO \leq 1.0$ Pressure $\leq$ 1500 psia Productivity $\geq$ 300 gMA/kg(cat)/hr EtOH Selectivity $\geq 75\%$ (CO_2-free)	Demonstrate 500 hours catalyst lifetime at 2009 performance. with biomass syngas for mixed alcohol catalyst cost of $\leq$ \$0.22/gal EtOH	$H_2/CO \leq 1.0$ Pressure $\leq$ 1000 psia Productivity $\geq$600 gMA/kg(cat)/hr EtOH Selectivity $\geq 80\%$ (CO_2- free)	Demonstrate 1000 hours catalyst lifetime at 2009 performance. with biomass syngas

Table 1. (Continued)

R&D Area	Current	2007	2008	2009	2010	2011	2012
Integration and Modeling	Research state-of-technology - 56 gal/dry ton EtOH $2.02/gal minimum EtOH selling price (higher alcohols sold at 85% of market value) at	Biomass Gasification/ Mixed Alcohol Design Report - Establishes a cost and quality baseline for technology	Improved hydrocarbon conversion efficiency yields- 56 gal/dry ton EtOH $1.73/gal minimum EtOH selling	Validated $1.73/gal EtOH for integrated Cleanup & Conditioning + Mixed Alcohol synthesis	Demonstrate feasibility of system (8000 hr on stream with ≤10% catalyst losses per year) based on regenerating	Validated $1.35/gal EtOH for integrated Cleanup & Condtioning + Mixed Alcohol synthesis	Demonstrate mixed alcohol yields of 89 gal/ton (76 gal/dry ton EtOH) via indirect biomass gasification at pilot-scale for "$1.07" minimum ETOH selling price
	$2.71/gal installed capital costs.	improvements for $1.07/gal thermochemica l ethanol by 2012 from indirect biomass gasification through a clean syngas intermediate.	price (higher alcohols priced as gasoline on an energy adjusted basis - $1.15/gal) at $2.69/gal installed capital costs.		fluidizable tar reforming catalyst to eliminate SMR		(higher alcohols priced as gasoline on an energy adjusted basis - $1.15/gal). Total installed capital costs are $2.31/annual gallon of ethanol.

Table 2. Tar Reformer Performance – % Conversion

Compound	Current (2005)	Goal (2012)
Methane (CH_4)	20%	80%
Ethane (C_2H_6)	90%	99%
Ethylene (C_2H_4)	50%	90%
Tars (C_{10}+)	95%	99.9%
Benzene (C_6H_6)	70%	99%
Ammonia (NH_3)	70%	90%

The current lab-scale demonstration results and target conversions for various impurities measured in biomass-derived syngas are listed in Table 2 for the year 2005 "current" state of technology case and the year 2012 "goal" case. The goal case conversions were selected to yield an economically viable clean syngas that is suitable for use in a catalytic fuel synthesis process without further hydrocarbon conversion steps.

The research target will be met when tar and light hydrocarbons are sufficiently converted to additional syngas, technically validating the elimination of a downstream steam methane reforming unit operation to separately reform methane from the other light hydrocarbons. Specific research to generate the required chemical and engineering data to design and successfully demonstrate a regenerating tar reforming reactor for long-term, reliable gas cleanup and conditioning includes:

- Performing tar deactivation/regeneration cycle tests to determine activity profiles to maintain the required long-term tar reforming catalyst activity
- Performing fundamental catalyst studies to determine deactivation kinetics and mechanisms by probing catalyst surfaces to uncover molecular-level details
- Determining optimized catalyst formulations and materials at the pilot scale to demonstrate catalyst performance and lifetimes as a function of process conditions and feedstock

Although consolidated tar and light hydrocarbon reforming tests performed with Ni-based catalysts have demonstrated the technical feasibility of this gas cleanup and conditioning strategy, alternative catalyst formulations can be developed to optimize reforming catalyst activity and lifetime in addition to expanded functionality. Specific further improvements that could be realized in catalyst functionality are:

- Further process intensification is possible by designing catalysts with higher tolerances for sulfur and chlorine poisons.
- Further reductions in gas cleanup costs could be realized by lowering or eliminating the sulfur and chlorine removal cost prior to reforming.
- Optimizing the water gas shift activity of reforming catalysts could reduce or eliminate the need for an additional downstream shift reactor.

Mixed Alcohol Synthesis R&D Needs

The ability to produce mixed alcohols from syngas has been known since the beginning of the last century; however, the commercial success of mixed alcohol synthesis has been limited by poor selectivity and low product yields. Single-pass yields are on the order of 10% syngas conversion (3 8.5% CO conversion) to alcohols, with methanol typically being the most abundant alcohol produced (Wender 1996; Herman 2000). For mixed alcohol synthesis to become an economical commercial process, there is a need for improved catalysts that increase the productivity and selectivity to higher alcohols (Fierro 1993).

Improvements in mixed alcohol synthesis catalysts could potentially increase alcohol yields and selectivity of ethanol production from clean syngas and improve the overall economics of the process through better heat integration and control and fewer syngas recycling loops. Specific research targets to achieve the cost-competitive 2012 target case are:

- Develop improved mixed alcohol catalysts that will increase the single-pass CO conversion from 3 8.5% to 50% and potentially higher and improve the CO selectivity to alcohols from 80% to 90%.
- Develop improved mixed alcohol catalysts with higher activity that will require a lower operating pressure (1,000 psia compared with 2,000 psia) to significantly lower process operating costs. This combination of lower syngas pressure for alcohol synthesis and less unconverted syngas to recompress and recycle has the added benefit of lowering the energy requirement for the improved synthesis loop.
- Alternative mixed alcohol synthesis reactors and catalysts should be explored. Greatly improved temperature control of the exothermic synthesis reaction has been demonstrated to significantly improve yields and product selectivity. Precise temperature control reactor designs need to be developed for the mixed alcohol synthesis reaction to improve the yields and the economics of the process.

Integration/Demonstration

As is the case for any sophisticated conversion process, combining the individual unit operations into a complete, integrated systematic process is a significant challenge. Individual pilot-scale operations to demonstrate the required performance of the unit operations as well as complete integrated pilot development runs will be required to demonstrate the cost-competitive technology. A specific challenge will be to continue to demonstrate process intensification and higher yields at pilot scale to reduce capital costs.

Achieving the technical target for the accelerated path to thermochemical ethanol requires meeting the specific research targets as outlined above. Missing or delaying any of these targets forfeits the 2012 target and jeopardizes the deployment of technologies in time to meet the 30x30 goal. The cost implications of missing, hitting or exceeding a target or set of targets are easily determined with process uncertainty analysis that will be performed and detailed in the upcoming Mixed Alcohol Design Report due in January 2007. Combinations of sensitivity analysis can provide several ways to achieve the same cost-competitive target, which reduces the overall risk of the process. Quantifying the relative cost savings for process improvements allows work to be directed to the most cost effective R&D to achieve the 2012 technical target for thermochemical ethanol production.

REFERENCES

Aden, A. & Spath,. P. L. (2005) Milestone Completion Report "*The Potential of Thermochemical Ethanol Via Mixed Alcohols Production.*" September 2005. http://devafdc.nrel.gov/bcfcdoc/9432.doc

Bizzari, S. N., Gubler, R. & Kishi, A. (2002). "*Oxo Chemicals.*" Chemical Economics Handbook, SRI International, Menlo Park, CA. Report number 682.7000

Foust, T. D., Wooley, R., Sheehan, J., Wallace, R., Ibsen, K., Dayton, D., Himmel, M., Ashworth, J., McCormick, R., Melendez, M., Hess, J. R., Kenney, K., Wang, M. & Snyder, S.; (2006). Werpy, T. *A National Laboratory Market and Technology Assessment of the 30X30 Scenario*. Draft NREL Report. December.

Fierro, J. L. G. (1993). "Catalysis in C1 chemistry: future and prospect." *Catalysis Letters 22(1- 2):* 67-91.

Herman, R. G. (2000). "Advances in catalytic synthesis and utilization of higher alcohols." *Catalysis Today 55(3)*: 233-245.

Wender, I. (1996). "Reactions of synthesis gas." *Fuel Processing Technology* 48(3): 189-297.

Spath, P. L. & Dayton, D. C. (2003). "Preliminary Screening -- Technical and Economic Assessment of Synthesis Gas to Fuels and Chemicals with Emphasis on the Potential for *Biomass-Derived Syngas.*" 160 pp.; NREL Report No. TP-510-34929.

APPENDIX C. NREL BIOREFINERY DESIGN DATABASE DESCRIPTION AND SUMMARY

NREL's Process Engineering Team has developed a database of primary information on all of the equipment in the benchmark model. This database contains information about the cost, reference year, scaling factor, scaling characteristic, design information and back-up cost referencing. The information is stored in a secure database and can be directly linked to the economic portion of the model. In addition to having all of the cost information used by the model, it has the ability to store documents pertaining to the piece of equipment. These include sizing and costing calculations and vendor information when available.

The following summarizes the important fields of information contained in the database. A partial listing of the information is attached for each piece of equipment. Additional information from the database is contained in the equipment cost listing in Appendix D.

Equipment Number:[AB]	Unique identifier, the first letter indicates the equipment typeand the first number represents the process area, e.g., P-301 is a pump in Area 300
Equipment Name:[AB]	Descriptive name of the piece of equipment
Associated PFD:	PFD number on which the piece of equipment appears, e.g., PFD-P800-A101
Equipment Category:[A]	Code indicating the general type of equipment, e.g., PUMP
Equipment Type:[A]	Code indicating the specific type of equipment, e.g., CENTRIFUGAL for a pump
Equipment Description:[A]	Short description of the size or characteristics of the piece of

	equipment, e.g., 20 gpm, 82 ft head for a pump
Number Required:[B]	Number of duplicate pieces of equipment needed
Number Spares:[B]	Number of on-line spares
Scaling Stream:[B]	Stream number or other characteristic variable from the ASPEN model by which the equipment cost will be scaled
Base Cost:[B]	Equipment cost
Cost Basis:[A]	Source of the equipment cost, e.g., ICARUS or VENDOR
Cost Year:[B]	Year for which the cost estimate is based
Base for Scaling:[B]	Value of the scaling stream or variable used to obtain the base cost of the equipment
Base Type:	Type of variable used for scaling, e.g., FLOW, DUTY, etc.
Base Units:	Units of the scaling stream or variable, e.g., KG/HR, CAL/S
Installation Factor:[B]	Value of the installation factor. Installed Cost = Base Cost x Installation Factor
Installation Factor Basis:	Source of the installation factor value, e.g., ICARUS, VENDOR
Scale Factor Exponent:[B]	Value of the exponential scaling equation
Scale Factor Basis:	Source of the scaling exponent value, e.g., GARRETT, VENDOR
Material of Construction:[A]	Material of Construction
Notes:	Any other important information about the design or cost
Document:	Complete, multi-page document containing design calculations, vendor literature and quotations and any other important information. This is stored as an electronic document and can be pages from a spreadsheet other electronic sources or scanned information from vendors.
Design Date:	Original date for the design of this piece of equipment
Modified Date:	The system automatically marks the date in this field any field is changed

[A] These fields are listed for all pieces of equipment in this Appendix.
[B] These fields are part of the equipment cost listing in Appendix D.

Table 2.

EQUIPMENT_NU	EQUIPMENT_NAME	EQUIPMENT _CATEGO	EQUIPMENT_TYPE	EQUIPMENT_DESCRIPTION	MATE-RIAL_ CONS	COST_BASIS
PFD-P800-A101-2						
C-101	Hopper Feeder	CONVEYOR	VIBRATING- FEEDER	Included in overall cost for feed handling & drying taken from several literature sources	CS	LITERATURE
C-102	Screener Feeder Conveyor	CONVEYOR	BELT	Included in overall cost for feed handling & drying taken from several literature sources	CS	LITERATURE
C-103	Radial Stacker Conveyor	CONVEYOR	BELT	Included in overall cost for feed handling & drying taken from several literature sources	CS	LITERATURE
C-104	Dryer Feed Screw Conveyor	CONVEYOR	SCREW	Included in overall cost for feed handling & drying taken from several literature sources	CS	LITERATURE
C-105	Gasifier Feed Screw Conveyor	CONVEYOR	SCREW	Included in overall cost for feed handling & drying taken from several literature sources	316SS	LITERATURE
H-286B	Flue Gas Cooler/Steam Generator #1	HEATX	SHELL-TUBE	duty = 155 MMBtu/hr; LMTD = 733 F; U = 150 Btu/hr-ft^2-F; area = 1,410 ft^2; fixed tube sheet	CS/INCL	QUESTIMATE
H-286C	Flue Gas Cooler /Boiler Water Preheater #1	HEATX	SHELL-TUBE	duty = 20 MMBtu/hr; LMTD = 244 F; U = 100 Btu/hr-ft^2-F; area = 823 ft^2; fixed TS	CS/A214	QUESTIMATE
H-31 1B	Flue Gas Cooler / Steam Generator #3	HEATX	SHELL-TUBE	duty = 47.9 MMBtu/hr; LMTD = 457; area = 698 sq ft; U = 150 Btu/hr-ft^2-F; fixed TS	CS/316S	ICARUS
K-101	Flue Gas Blower	FAN	CENTRIFUGAL	Included in overall cost for feed handling & drying taken from several literature sources	SS304	LITERATURE
M-101	Hydraulic Truck Dump with Scale	SCALE	TRUCK-SCALE	Included in overall cost for feed handling & drying taken from several literature sources		LITERATURE
M-102	Hammermill	SIZE-REDUCTION		Included in overall cost for feed handling & drying taken from several literature sources	CS	LITERATURE
M-103	Front End Loaders	VEHICLE	LOADER	Included in overall cost for feed handling & drying taken from several literature sources	CS	LITERATURE

(Continued)

EQUIPMENT_NU	EQUIPMENT_NAME	EQUIPMENT _CATEGO	EQUIPMENT_TYPE	EQUIPMENT_DESCRIPTION	MATE-RIAL_ CONS	COST_BASIS
M-104	Rotary Biomass Dryer	DRYER	ROTARY-DRUM	Included in overall cost for feed handling & drying taken from several literature sources	CS	LITERATURE
S-101	Magnetic Head Pulley	SEPARATOR	MAGNET	Included in overall cost for feed handling & drying taken from several literature sources	CS	LITERATURE
S-102	Screener	SEPARATOR	SCREEN	Included in overall cost for feed handling & drying taken from several literature sources	CS	LITERATURE
S-103	Dryer Air Cyclone	SEPARATOR	GAS CYCLONE	Included in overall cost for feed handling & drying taken from several literature sources	CS	LITERATURE
S-104	Dryer Air Baghouse Filter	SEPARATOR	FABRIC-FILTER	Included in overall cost for feed handling & drying taken from several literature sources		LITERATURE
T-1 01	Dump Hopper	TANK	LIVE-BTM-BIN	Included in overall cost for feed handling & drying taken from several literature sources	CS	LITERATURE
T-1 02	Hammermill Surge Bin	TANK	LIVE-BTM-BIN	Included in overall cost for feed handling & drying taken from several literature sources	CS	LITERATURE
T-1 03	Dryer Feed Bin	TANK	LIVE-BTM-BIN	Included in overall cost for feed handling & drying taken from several literature sources	CS	LITERATURE
T-1 04	Dried Biomass Hopper	TANK	VERTICAL-VESSEL	Included in overall cost for feed handling & drying taken from several literature sources	CS	LITERATURE
PFD-P800-A201						
C-201	Sand/ash Conditioner/Conveyor	CONVEYOR	SCREW	Included in overall cost for gasification & gas clean up taken from several literature sources	CS	LITERATURE
K-202	Combustion Air Blower	FAN	CENTRIFUGAL	Included in overall cost for gasification & gas clean up taken from several literature sources	CS	LITERATURE

(Continued)

EQUIPMENT_NU	EQUIPMENT_NAME	EQUIPMENT_CATEGO	EQUIPMENT_TYPE	EQUIPMENT_DESCRIPTION	MATE-RIAL_CONS	COST_BASIS
M-201	Sand/ash Cooler	MISCELLANEOUS	MISCELLANEOUS	Included in overall cost for gasification & gas clean up taken from several literature sources		LITERATURE
R-201	Indirectly-heated Biomass Gasifier	REACTOR	VERTICAL-VESSEL	Included in overall cost for gasification & gas clean up taken from several literature sources	CS w/refract-ory	LITERATURE
R-202	Char Combustor	REACTOR	VERTICAL-VESSEL	Included in overall cost for gasification & gas clean up taken from several literature sources	CS w/refract-ory	LITERATURE
S-201	Primary Gasifier Cyclone	SEPARATOR	GAS CYCLONE	Included in overall cost for gasification & gas clean up taken from several literature sources	CS w/refract-ory	LITERATURE
S-202	Secondary Gasifier Cyclone	SEPARATOR	GAS CYCLONE	Included in overall cost for gasification & gas clean up taken from several literature sources	CS w/refract-ory	LITERATURE
S-203	Primary Combustor Cyclone	SEPARATOR	GAS CYCLONE	Included in overall cost for gasification & gas clean up taken from several literature sources	CS w/refract-ory	LITERATURE
S-204	Secondary Combustor Cyclone	SEPARATOR	GAS CYCLONE	Included in overall cost for gasification & gas clean up taken from several literature sources	CS w/refract-ory	LITERATURE
S-205	Electrostatic Precipitator	SEPARATOR	MISCELLANEOUS	Included in overall cost for gasification & gas clean up taken from several literature sources	CS	LITERATURE
T-201	Sand/ash Bin	TANK	FLAT-BTMSTORAGE	Included in overall cost for gasification & gas clean up taken from several literature sources	CS	LITERATURE
PFD-P800-A301-5						
H-301	Quench Water Recirculation Cooler	HEATX	SHELL-TUBE	Included in overall cost for gasification & gas clean up taken from several literature sources	CS	LITERATURE
H-301A	Post-tar Reformer Cooler / Steam Generator #2	HEATX	SHELL-TUBE	duty = 47.9 MMBtu/hr; LMTD = 457; area = 698 sq ft; U = 150 Btu/hr-ft^2-F; fixed TS	CS/316S	ICARUS

(Continued)

EQUIPMENT_NU	EQUIPMENT_NAME	EQUIPMENT _CATEGO	EQUIPMENT_TYPE	EQUIPMENT_DESCRIPTION	MATE-RIAL_ CONS	COST_BASIS
H-301B	Reformer Flue Gas Cooler/Steam superheater Reformed Syngas / Synthesis Reactor Preheat cooler	HEATX	SHELL-TUBE	duty = 94 MMBtu/hr; LMTD = 217 F; U = 150 Btu/hr-ft^2-F; area = 2,900 ft^2; fixed TS	CS/INCL	QUESTIMATE
H-301C	#1	HEATX	SHELL-TUBE	duty =40.0 MMBtu/hr; LMTD = ?? F; U = 90 Btu/hr-ft^2-F; surface area = 1,552 ft^2	A214	QUESTIMATE
H-302	Syngas Compressor Intercoolers	HEATX	AIR-COOLED EXCHANGER	Cost of intercoolers included in cost for syngas compressor, K-301	CS	ICARUS
H-303	Water-cooled Aftercooler	HEATX	SHELL-TUBE	duty = 2.9 MMBtu/hr; LMTD = 25F; U = 150 Btu/hr-ft^2-F; surface area = 794 ft^2; fixed TS	SS304CS /A214	QUESTIMATE
H-304	LO-CAT Preheater	HEATX	SHELL-TUBE	duty = 0.8 MMBtu/hr;LMTD = 87 F; U = 90 Btu/hr-ft^2-F; surface area = 98 ft^2; fixed TS	A285C/C A443	QUESTIMATE
H-305	LO-CAT Absorbent Solution Cooler	HEATX	SHELL-TUBE	Included in LO-CAT system cost	304SS	VENDOR
H-315D1	Recycle Syngas Cooler / Steam Generator #4	HEATX	SHELL-TUBE	duty = 1.37 MMBtu/hr; LMTD = 1,220 F; U = 150 Btu/hr-ft^2-F; area = 7 ft^2; fixed TS	CS/INCL	QUESTIMATE
H-315D2	Recycle Syngas cooler #2 / Air preheat #1	HEATX	SHELL-TUBE	duty = 0.8 MMBtu/hr;LMTD = 87 F; U = 90 Btu/hr-ft^2-F; surface area = 98 ft^2; fixed TS	A285C/C A443	QUESTIMATE
K-301	Syngas Compressor	COMPRESSOR	CENTRIFUGAL	gas flow rate = 70,000 CFM; 6 impellers; design outlet pressure = 465 psi; 30,000 HP; intercoolers, aftercooler, & K.O.s included	A285C	QUESTIMATE
K-302	LO-CAT Feed Air Blower	FAN	CENTRIFUGAL	Included in LO-CAT system cost	CS	VENDOR
K-305	Regenerator Combustion Air Blower	FAN	CENTRIFUGAL	gas flow rate (actual) = 70133 CFM;	SS304	QUESTIMATE

(Continued)

EQUIPMENT_NU	EQUIPMENT_NAME	EQUIPMENT_CATEGO	EQUIPMENT_TYPE	EQUIPMENT_DESCRIPTION	MATE-RIAL_CONS	COST_BASIS
M-301	Syngas Quench Chamber	MISCELLANEOUS		Included in overall cost for gasification & gas clean up taken from several literature sources	CS	LITERATURE
M-302	Syngas Venturi Scrubber	MISCELLANEOUS		Included in overall cost for gasification & gas clean up taken from several literature sources	CS	LITERATURE
M-303	LO-CAT Venturi Precontactor	MISCELLANEOUS		Included in LO-CAT system cost	304SS	VENDOR
M-304	LO-CAT Liquid-filled Absorber	COLUMN	ABSORBER	Included in LO-CAT system cost	304SS	VENDOR
P-301	Sludge Pump	PUMP	CENTRIFUGAL	1.4 GPM; 0.053 brake HP; design pressure = 60 psia	CS	QUESTIMATE
P-302	Quench Water Recirculation Pump	PUMP	CENTRIFUGAL	Included in the cost of the gasification & gas clean up system	CS	LITERATURE
P-303	LO-CAT Absorbent Solution Circulating Pump	PUMP	CENTRIFUGAL	Included in LO-CAT system cost	304SS	VENDOR
R-301A	Tar Reformer Catalyst Regenerator	REACTOR	VERTICAL-VESSEL	Taken from literature source	CS w/refract ory	LITERATURE
R-303	Tar Reformer	REACTOR	VERTICAL-VESSEL	Included in overall cost for gasification & gas clean up taken from several literature sources	CS w/refract ory	LITERATURE
R-304	LO-CAT Oxidizer Vessel	REACTOR	VERTICAL-VESSEL	Included in LO-CAT system cost	304SS	VENDOR
S-301	Pre-compressor Knock-out	SEPARATOR	KNOCK-OUT DRUM	18 ft diameter; 36 ft height; design pres = 40 psia; design temp = 197 F	CS	QUESTIMATE
S-302	Syngas Compressor Interstage Knock-outs	SEPARATOR	KNOCK-OUT DRUM	Cost of intercoolers K.O.s included in cost for syngas compressor, K-301	CS	ICARUS
S-303	Post-compressor Knock-out	SEPARATOR	KNOCK-OUT DRUM	7 ft. diameter; 14 ft height; design pres = 506 psia; design temp = 160 F	CS	QUESTIMATE
S-306	Tar Reformer Cyclone	SEPARATOR	GAS CYCLONE	Included in the cost of the tar reformer catalyst renegerator, R-204	CS	LITERATURE
S-307	Catalyst Regenerator Cyclone	SEPARATOR	GAS CYCLONE	Included in the cost of the tar reformer catalyst renegerator, R-204	CS	LITERATURE

(Continued)

EQUIPMENT_NU	EQUIPMENT_NAME	EQUIPMENT _CATEGO	EQUIPMENT_TYPE	EQUIPMENT_DESCRIPTION	MATE-RIAL_ CONS	COST_BASIS
S-310	L.P. Amine System	COLUMN	ABSORBER			OTHER
T-301	Sludge Settling Tank	SEPARATOR	CLARIFIER	3 ft diameter; 7 ft height; 431 gal volume;	SS304	QUESTIMATE
T-302	Quench Water Recirculation Tank	TANK	HORIZONTAL-VESSEL	Included in overall cost for gasification & gas clean up taken from several literature sources	CS	LITERATURE
PFD-P800-A401-2						
H-410B	Flue Gas Cool / syngas rxn preheat	HEATX	SHELL-TUBE	duty = 0.8 MMBtu/hr;LMTD = 87 F; U = 90 Btu/hr-ft^2-F; surface area = 98 ft^2; fixed TS	A285C/C A443	QUESTIMATE
H-41 1A	Air preheat #3 / post Reactor Syngas cooling #1 Post synthesis cooler #2/Deaerator Water Preheater	HEATX	SHELL-TUBE	duty = 0.8 MMBtu/hr;LMTD = 87 F; U = 90 Btu/hr-ft^2-F; surface area = 98 ft^2; fixed TS	A285C/C A443	QUESTIMATE
H-41 1 B	#1	HEATX	SHELL-TUBE	duty = 20 MMBtu/hr; LMTD = 244 F; U = 100 Btu/hr-ft^2-F; area = 823 ft^2; fixed TS	CS/A214	QUESTIMATE
H-41 1 C	Post Synthesis cooler #3/Makeup Water heater	HEATX	SHELL-TUBE	duty = 20 MMBtu/hr; LMTD = 244 F; U = 100 Btu/hr-ft^2-F; area = 823 ft^2; fixed TS	CS/A214	QUESTIMATE
H-41 1D	Post Synthesis cooler #4 / syngas recycle heat #1	HEATX	SHELL-TUBE	duty = 0.8 MMBtu/hr;LMTD = 87 F; U = 90 Btu/hr-ft^2-F; surface area = 98 ft^2; fixed TS	A285C/C A443	QUESTIMATE
H-41 1E	Post Synthesis Cooler #5/Mol Sieve preheater #2	HEATX	SHELL-TUBE	duty = 20 MMBtu/hr; LMTD = 244 F; U = 100 Btu/hr-ft^2-F; area = 823 ft^2; fixed TS	CS/A214	QUESTIMATE
H-412	Post Mixed Alcohol Cooler	HEATX	SHELL-TUBE	duty = 76.5 MMBtu/hr; LMTD = ?? F; U = 100 Btu/hr-ft^2-F; surface area = 4,763 ft^2	A214	QUESTIMATE
H-41 3	Mixed Alcohol first Condenser (air cooled)	HEATX	AIR-COOLED EXCHANGER			
H-414	Mixed Alcohol Condenser	HEATX	SHELL-TUBE	duty = 78 MMBtu/hr; LMTD = 41 F; U = 150 Btu/hr-ft^2-F; surface area = 12,462 ft^2	A214	QUESTIMATE

(Continued)

EQUIPMENT_NU	EQUIPMENT_NAME	EQUIPMENT _CATEGO	EQUIPMENT_TYPE	EQUIPMENT_DESCRIPTION	MATE-RIAL_ CONS	COST_BASIS
H-416B	Recycle Syngas Heat #2 / Flue gas Cool	HEATX	SHELL-TUBE	duty =40.0 MMBtu/hr; LMTD = ?? F; U = 90 Btu/hr-ft^2-F; surface area = 1,552 ft^2	A214	QUESTIMATE
K-410	Mixed Alcohol Gas Compressor	COMPRESSOR	CENTRIFUGAL	gas flow rate = 2,481 CFM; 4 impellers; design outlet pressure = 700 psi; 10,617 HP; intercoolers, aftercooler, & K.O.s included	A285C	QUESTIMATE
K-412	Purge Gas Expander	COMPRESSOR	CENTRIFUGAL	gas flow rate = 144 CFM; design outlet pressure =25 psi; 2740 HP	A285C	QUESTIMATE
R-41 0	Mixed Alcohol Reactor	REACTOR	VERTICAL-VESSEL	Fixed Bed Synthesis Reactor with MoS2-based catalyst. Sized from hourly space velocity of 3000	CS w/refract ory	QUESTIMATE
S-501	Mixed Alcohols Condensation Knock-out	SEPARATOR	KNOCK-OUT DRUM	H/D = 2; 5 ft diam; 9 ft height; operating pressure = 1993 psia; operating temperature = 110 F	A-515	QUESTIMATE
PFD-P800-A501-2						
D-504	Ethanol/Propanol Splitter	COLUMN	DISTILLATION			
D-505	Methanol/Ethanol Splitter	COLUMN	DISTILLATION			
H-503A	Syngas Cooler #4 / Mol Sieve preheater #1	HEATX	SHELL-TUBE	duty = 20 MMBtu/hr; LMTD = 244 F; U = 100 Btu/hr-ft^2-F; area = 823 ft^2; fixed TS	CS/A214	QUESTIMATE
H-503B	Mol Sieve Superheater / reformed syngas cool #5	HEATX	SHELL-TUBE	duty = 0.8 MMBtu/hr;LMTD = 87 F; U = 90 Btu/hr-ft^2-F; surface area = 98 ft^2; fixed TS	A285C/C A443	QUESTIMATE
H-504C	D-504 condenser (air cooled)	HEATX	AIR-COOLED EXCHANGER			
H-504R	Ethanol/Propanol Splitter Reboiler	HEATX	SHELL-TUBE			
H-505C	D-505 condenser (air cooled)	HEATX	AIR-COOLED EXCHANGER			
H-505R	Methanol/Ethanol Splitter Reboiler	HEATX	SHELL-TUBE			

(Continued)

EQUIPMENT_NU	EQUIPMENT_NAME	EQUIPMENT _CATEGO	EQUIPMENT_TYPE	EQUIPMENT_DESCRIPTION	MATE-RIAL_ CONS	COST_BASIS
H-51 3	Mol Sieve Flush Condenser (air cooled)	HEATX	AIR-COOLED EXCHANGER			
H-590	MA Product Cooler / Mol Sieve preheater #3	HEATX	SHELL-TUBE	duty = 3 MMBtu/hr; LMTD = 236 F; U = 600 Btu/hr-ft^2-F; area = 20 ft^2; pre-engineered U-tube	A285C/C A443	QUESTIMATE
H-591	Higher Alcohol Product Finishing cooler	HEATX	SHELL-TUBE	Included in overall cost for gasification & gas clean up taken from several literature sources	CS	LITERATURE
H-592	ETHANOL Product Cooler / Mol Sieve preheater #4	HEATX	SHELL-TUBE	duty = 3 MMBtu/hr; LMTD = 236 F; U = 600 Btu/hr-ft^2-F; area = 20 ft^2; pre-engineered U-tube	A285C/C A443	QUESTIMATE
H-593	ETHANOL Product Finishing cooler	HEATX	SHELL-TUBE	Included in overall cost for gasification & gas clean up taken from several literature sources	CS	LITERATURE
P-504B	Ethanol/Propanol Splitter Bottoms Pump	PUMP	CENTRIFUGAL			
P-504R	Ethanol/Propanol Splitter Reflux Pump	PUMP	CENTRIFUGAL			
P-505B	Methanol/Ethanol Splitter Bottoms Pump	PUMP	CENTRIFUGAL			
P-505R	Methanol/Ethanol Splitter Reflux Pump	PUMP	CENTRIFUGAL			
S-502	Methanol Separation Column	COLUMN	DISTILLATION	13.5' dia, 32 Actual Trays, Nutter V-Grid Trays	SS304	ICARUS
S-503	Molecular Sieve (9 pieces)	MISCELLANEOUS	PACKAGE	Superheater, twin mole sieve columns, product cooler, condenser, pumps, vacuum source.	SS	VENDOR
T-504	Ethanol/Propanol Splitter Reflux Drum	TANK	KNOCK-OUT DRUM			
T-505	Methanol/Ethanol Splitter Reflux Drum	TANK	KNOCK-OUT DRUM			
T-590	Mixed Alcohol Product Storage Tank	TANK	FLAT-BTMSTORAGE			
T-592	Ethanol Product Storage Tank	TANK	FLAT-BTMSTORAGE			

(Continued)

EQUIPMENT_NU	EQUIPMENT_NAME	EQUIPMENT _CATEGO	EQUIPMENT_TYPE	EQUIPMENT_DESCRIPTION	MATE-RIAL_ CONS	COST_BASIS
PFD-P800-A601-3						
H-601	Steam Turbine Condenser	HEATX	SHELL-TUBE	Included in the cost of the steam trubine/generator (M-602); condenser steam flow rate = 342,283 lb/hr		ADEN, ET. AL. 2002
H-602	Blowdown Cooler / Deaerator Water Preheater	HEATX	SHELL-TUBE	duty = 3 MMBtu/hr; LMTD = 236 F; U = 600 Btu/hr-ft^2-F; area = 20 ft^2; pre-engineered U-tube	A285C/C A443	QUESTIMATE
H-603	Blowdown Water-cooled Cooler	HEATX	SHELL-TUBE	duty = 0.6 MMBtu/hr; LMTD = 47 F; U = 225 Btu/hr-ft^2-F; area = 60 ft^2; fixed TS	A214	QUESTIMATE
M-601	Hot Process Water Softener System	MISCELLANEOUS	PACKAGE	scaled cost to 700 gpm flow, 24" dia softener. Includes filters, chemical feeders, piping, valves		RICHARDSON
M-602	Extraction Steam Turbine/Generator	GENERATOR	STEAM-TURBINE	25.6 MW generated; 34,308 HP		VENDOR
M-603	Startup Boiler	MISCELLANEOUS	PACKAGE	Assume need steam requirement equal to 1/2 of steam requirement for gasifier at full rate steam rate = 36,560 lb/hr	CS	QUESTIMATE
P-601	Collection Pump	PUMP	CENTRIFUGAL	513 GPM; 4 brake HP; outlet pressure = 25 psia	CS	QUESTIMATE
P-602	Condensate Pump	PUMP	CENTRIFUGAL	190 GPM; 4 brake HP; outlet pressure = 25 psia	SS304	QUESTIMATE
P-603	Deaerator Feed Pump	PUMP	CENTRIFUGAL	702 GPM; 14 brake HP; outlet pressure = 40 psia	CS	QUESTIMATE
P-604	Boiler Feed Water Pump	PUMP	CENTRIFUGAL	730 GPM; 759 brake HP; outlet pressure = 1,345 psia	CS	QUESTIMATE
S-601	Blowdown Flash Drum	TANK	HORIZONTAL-VESSEL	H/D = 2; residence time = 5 min; 2 ft diameter; 4 ft height; op press = 1,280 psi; op temp = 575 F	CS	QUESTIMATE
T-601	Condensate Collection Tank	TANK	HORIZONTAL-VESSEL	residence time = 10 minutes; H/D = 2; 8 ft diameter; 17 ft height	CS	QUESTIMATE
T-602	Condensate Surge Drum	TANK	HORIZONTAL-VESSEL	residence time = 10 minutes; H/D = 2; 9 ft diameter; 17 ft height	CS	QUESTIMATE

(Continued)

EQUIPMENT_NU	EQUIPMENT_NAME	EQUIPMENT _CATEGO	EQUIPMENT_TYPE	EQUIPMENT_DESCRIPTION	MATE-RIAL_ CONS	COST_BASIS
T-603	Deaerator	TANK	HORIZONTAL-VESSEL	liquid flow rate = 348,266 lb/hr; 150 psig design pressure; 10 min residence time	CS;SS316	VENDOR
T-604	Steam Drum	TANK	HORIZONTAL-VESSEL	424 gal, 4.5' x 4'dia, 15 psig	CS	ICARUS
PFD-P800-A701-2						
K-701	Plant Air Compressor	COMPRESSOR	RECIPROCATING	450 cfm, 125 psig outlet	CS	ICARUS
M-701	Cooling Tower System	COOLING-TOWER	INDUCED-DRAFT	approx 16,500 gpm, 140 MMBtu/hr	FIBERGLASS	DELTA-T98
M-702	Hydraulic Truck Dump with Scale	SCALE	TRUCK-SCALE	Hydraulic Truck Dumper with Scale	CS	VENDOR
M-703	Flue Gas Stack	MISCELLANEOUS	MISCELLANEOUS	42 inch diameter; 250 deg F	A515	QUESTIMATE
P-701	Cooling Water Pump	PUMP	CENTRIFUGAL	16,188 GPM; 659 brake HP; outlet pressure 75 psi	CS	QUESTIMATE
P-702	Firewater Pump	PUMP	CENTRIFUGAL	2,500 gpm, 50 ft head	CS	ICARUS
P-703	Diesel Pump	PUMP	CENTRIFUGAL	30 gpm, 150 ft head	CS	ICARUS
P-704	Ammonia Pump	PUMP	CENTRIFUGAL	8.5 gpm, 22 ft head	CS	ICARUS
P-705	Hydrazine Pump	PUMP	CENTRIFUGAL	5 gpm, 75 ft head	CS	DELTA-T98
S-701	Instrument Air Dryer	DRYER	PACKAGE	400 SCFM Air Dryer, -40 F Dewpoint	CS	RICHARDSON
T-701	Plant Air Receiver	TANK	HORIZONTAL-VESSEL	900 gal., 200 psig	CS	ICARUS
T-702	Firewater Storage Tank	TANK	FLAT-BTMSTORAGE	600,000 gal, 4 hr res time, 51' dia x 40' high, atmospheric	A285C	ICARUS
T-703	Diesel Storage Tank	TANK	FLAT-BTMSTORAGE	10,667 gal, 120 hr res time, 90% wv, 10' dia x 18.2' high, atmospheric	A285C	ICARUS
T-704	Ammonia Storage Tank	TANK	HORIZONTAL-STORAGE	Included in the cost of the feed handling step.	A51 5	ICARUS
T-705	Olivine Lock Hopper	TANK	VERTICAL-VESSEL	Included in the cost of the feed handling step.	CS	DELTA-T98
T-706	MgO Lock Hopper	TANK	VERTICAL-VESSEL	20' x 20' Bin, Tapering to 3' x 3' at Bottom. Capacity 6,345 cf, two truck loads.	CS	DELTA-T98
T-707	Hydrazine Storage Tank	TANK	VERTICAL-VESSEL	260 gal, 4.9' x 3'dia., 10psig	SS316	ICARUS

APPENDIX D. INDIVIDUAL EQUIPMENT COST SUMMARY

Equipment Number	Number Required	Number Spares	Equipment Name	Size Ratio	Original Equip Cost (per unit)	Base Year	Total Original Equip Cost (Req'd & Spare) in Base Year	Scaling Exponent	Scaled Cost in Base Year	Installation Factor	Installed Cost in Base Year	Installed Cost in 2005$
C-101	4		Hopper Feeder	1.00	$0	2002	$0	0.75	$0	2.47	$0	$0
C-102	2		Screener Feeder Conveyor	1.00	$0	2002	$0	0.75	$0	2.47	$0	$0
C-103	2		Radial Stacker Conveyor	1.00	$0	2002	$0	0.75	$0	2.47	$0	$0
C-104	2		Dryer Feed Screw Conveyor	1.00	$0	2002	$0	0.75	$0	2.47	$0	$0
C-105	2		Gasifier Feed Screw Conveyor	0.93	$0	2002	$0	0.75	$0	2.47	$0	$0
H-286B	1		Flue Gas Cooler/Steam Generator #1	0.60	$347,989	2002	$347,989	0.6	$256,890	2.47	$634,519	$750,965
H-286C	1		Flue Gas Cooler /Boiler Water Preheater #1	0.03	$20,989	2002	$20,989	0.6	$2,637	2.47	$6,512	$7,708
H-311B	1		Flue Gas Cooler / Steam Generator #3	0.69	$69,089	2002	$69,089	0.65	$54,035	2.47	$133,465	$157,959
K-101	2		Flue Gas Blower	1.56	$0	2002	$0	0.75	$0	2.47	$0	$0
M-101	4		Hydraulic Truck Dump with Scale	1.00	$0	2002	$0	0.75	$0	2.47	$0	$0
M-102	2		Hammermill	1.00	$0	2002	$0	0.75	$0	2.47	$0	$0
M-103	3		Front End Loaders	1.00	$0	2002	$0	0.75	$0	2.47	$0	$0
M-104	2		Rotary Biomass Dryer	1.00	$3,813,728	2002	$7,627,455	0.75	$7,627,450	2.47	$18,839,801	$22,297,257
S-101	2		Magnetic Head Pulley	1.00	$0	2002	$0	0.75	$0	2.47	$0	$0
S-102	2		Screener	1.00	$0	2002	$0	0.75	$0	2.47	$0	$0
S-103	2		Dryer Air Cyclone	1.56	$0	2002	$0	0.75	$0	2.47	$0	$0
S-104	2		Dryer Air Baghouse Filter	0.93	$0	2002	$0	0.75	$0	2.47	$0	$0
T-101	4		Dump Hopper	1.00	$0	2002	$0	0.75	$0	2.47	$0	$0
T-102	1		Hammermill Surge Bin	1.00	$0	2002	$0	0.75	$0	2.47	$0	$0
T-103	2		Dryer Feed Bin	1.00	$0	2002	$0	0.75	$0	2.47	$0	$0

(Continued)

Equipment Number	Number Required	Number Spares	Equipment Name	Size Ratio	Original Equip Cost (per unit)	Base Year	Total Original Equip Cost (Req'd & Spare) in Base Year	Scaling Exponent	Scaled Cost in Base Year	Installation Factor	Installed Cost in Base Year	Installed Cost in 2005$
T-104	2		Dried Biomass Hopper	0.93	$0	2002	$0	0.75	$0	2.47	$0	$0
A100						**Subtotal**	**$8,065,522**		**$7,941,011**	**2.47**	**$19,614,298**	**$23,213,888**
C-201	1		Sand/ash Conditioner/Conveyor	0.33	$0	2002	$0	0.65	$0	2.47	$0	$0
K-202	2		Combustion Air Blower	0.97	$0	2002	$0	0.65	$0	2.47	$0	$0
M-201	2		Sand/ash Cooler	0.33	$0	2002	$0	0.65	$0	2.47	$0	$0
R-201	2		Indirectly-heated Biomass Gasifier	1.00	$2,212,201	2002	$4,424,402	0.65	$4,418,389	2.47	$10,913,421	$12,916,238
R-202	2		Char Combustor	1.00	$0	2002	$0	0.65	$0	2.47	$0	$0
S-201	2		Primary Gasifier Cyclone	1.00	$0	2002	$0	0.65	$0	2.47	$0	$0
S-202	2		Secondary Gasifier Cyclone	0.97	$0	2002	$0	0.65	$0	2.47	$0	$0
S-203	2		Primary Combustor Cyclone	1.00	$0	2002	$0	0.65	$0	2.47	$0	$0
S-204	2		Secondary Combustor Cyclone	0.95	$0	2002	$0	0.65	$0	2.47	$0	$0
S-205	2		Electrostatic Precipitator	0.96	$0	2002	$0	0.65	$0	2.47	$0	$0
T-201	1		Sand/ash Bin	0.33	$0	2002	$0	0.65	$0	2.47	$0	$0
A200						**Subtotal**	**$4,424,402**		**$4,418,389**	**2.47**	**$10,913,421**	**$12,916,238**
H-301	1		Quench Water Recirculation Cooler	1.57	$0	2002	$0	0.44	$0	2.47	$0	$0
H-301A	1		Post-tar Reformer Cooler / Steam Generator #2	1.66	$69,089	2002	$69,089	0.65	$96,054	2.47	$237,253	$280,793
H-301B	1		Reformer Flue Gas Cooler/Steam superheater	1.00	$196,589	2002	$196,589	0.6	$196,056	2.47	$484,259	$573,129

(Continued)

Equipment Number	Number Required	Number Spares	Equipment Name	Size Ratio	Original Equip Cost (per unit)	Base Year	Total Original Equip Cost (Req'd & Spare) in Base Year	Scaling Exponent	Scaled Cost in Base Year	Installation Factor	Installed Cost in Base Year	Installed Cost in 2005$
H-301C	1		Reformed Syngas cooler / Synthesis Reactor Preheat #1	0.32	$144,006	2002	$144,006	0.44	$87,219	2.47	$215,431	$254,966
H-302	5		Syngas Compressor Intercoolers	1.57	$0	2002	$0	0.65	$0	2.47	$0	$0
H-303	1		Water-cooled Aftercooler	1.81	$20,889	2002	$20,889	0.44	$27,111	2.47	$66,965	$79,254
H-304	1		LO-CAT Preheater	0.15	$4,743	2002	$4,743	0.6	$1,539	2.47	$3,800	$4,498
H-305	1		LO-CAT Absorbent Solution Cooler	0.29	$0	2002	$0	0.44	$0	2.47	$0	$0
H-315D1	1		Recycle Syngas Cooler / Steam Generator #4	4.20	$26,143	2002	$26,143	0.6	$61,841	2.47	$152,746	$180,778
H-315D2	1		Recycle Syngas cooler #2 / Air preheat #1	4.08	$4,743	2002	$4,743	0.6	$11,020	2.47	$27,219	$32,214
K-301	1		Syngas Compressor	1.73	$3,896,834	2002	$3,896,834	0.8	$6,036,915	2.47	$14,911,181	$17,647,662
K-302	1		LO-CAT Feed Air Blower	0.73	$0	2002	$0	0.65	$0	2.47	$0	$0
K-305	1		Regenerator Combustion Air Blower	0.94	$35,020	2002	$35,020	0.59	$33,806	2.47	$83,500	$98,824
M-301	1		Syngas Quench Chamber	1.57	$0	2002	$0	0.65	$0	2.47	$0	$0
M-302	1		Syngas Venturi Scrubber	1.57	$0	2002	$0	0.65	$0	2.47	$0	$0
M-303	1		LO-CAT Venturi Precontactor	0.73	$0	2002	$0	0.65	$0	2.47	$0	$0
M-304	1		LO-CAT Liquid-filled Absorber	0.29	$0	2002	$0	0.65	$0	2.47	$0	$0
P-301	1	1	Sludge Pump	0.08	$3,911	2002	$7,822	0.33	$3,351	2.47	$8,277	$9,795
P-302	1	1	Quench Water Recirculation Pump	0.99	$0	2002	$0	0.65	$0	2.47	$0	$0
P-303	1	1	LO-CAT Absorbent Solution Circulating Pump	1.57	$0	2002	$0	0.65	$0	2.47	$0	$0

(Continued)

Equipment Number	Number Required	Number Spares	Equipment Name	Size Ratio	Original Equip Cost (per unit)	Base Year	Total Original Equip Cost (Req'd & Spare) in Base Year	Scaling Exponent	Scaled Cost in Base Year	Installation Factor	Installed Cost in Base Year	Installed Cost in 2005$
R-301A	1		Tar Reformer Catalyst Regenerator	1.62	$2,429,379	2002	$2,429,379	0.65	$3,324,994	2.47	$8,212,736	$9,719,926
R-303	1		Tar Reformer	1.57	$2,212,201	2002	$2,212,201	0.65	$2,965,912	2.47	$7,325,802	$8,670,224
R-304	1		LO-CAT Oxidizer Vessel	0.73	$1,000,000	2002	$1,000,000	0.65	$813,486	2.47	$2,009,311	$2,378,057
S-301	1		Pre-compressor Knock-out	1.73	$157,277	2002	$157,277	0.6	$218,395	2.47	$539,436	$638,432
S-302	4		Syngas Compressor Interstage Knock-outs	1.73	$0	2002	$0	0.6	$0	2.47	$0	$0
S-303	1		Post-compressor Knock-out	1.86	$40,244	2002	$40,244	0.6	$58,421	2.47	$144,300	$170,781
S-306	1		Tar Reformer Cyclone	1.57	$0	2002	$0	0.65	$0	2.47	$0	$0
S-307	1		Catalyst Regenerator Cyclone	1.62	$0	2002	$0	0.65	$0	2.47	$0	$0
S-310	1		L.P. Amine System	1.26	$3,485,685	2002	$3,485,685	0.75	$4,155,524	2.47	$10,264,143	$12,147,805
T-301	1		Sludge Settling Tank	0.00	$11,677	2002	$11,677	0.6	$260	2.47	$641	$759
T-302	1		Quench Water Recirculation Tank	1.57	$0	2002	$0	0.65	$0	2.47	$0	$0
A300						**Subtotal**	**$13,42,341**		**$18,091,903**	2.47	**$44,687,000**	**$52,887,900**
H-410B	1		Flue Gas Cool / syngas rxn preheat	6.78	$4,743	2002	$4,743	0.6	$14,951	2.47	$36,928	$43,705
H-411A	1		Air preheat #3 / post Reactor Syngas cooling #1	2.65	$4,743	2002	$4,743	0.6	$8,505	2.47	$21,007	$24,862
H-411B	1		Post synthesis cooler #2/Deaerator Water Preheater #1	0.77	$20,989	2002	$20,989	0.6	$17,979	2.47	$44,409	$52,558
H-411C	1		Post Synthesis cooler #3/Makeup Water heater	0.33	$20,989	2002	$20,989	0.6	$10,757	2.47	$26,569	$31,445

(Continued)

Equipment Number	Number Required	Number Spares	Equipment Name	Size Ratio	Original Equip Cost (per unit)	Base Year	Total Original Equip Cost (Req'd & Spare) in Base Year	Scaling Exponent	Scaled Cost in Base Year	Installation Factor	Installed Cost in Base Year	Installed Cost in 2005$
H-411D	1		Post Synthesis cooler #4 / syngas recycle heat #1	5.20	$4,743	2002	$4,743	0.6	$12,758	2.47	$31,512	$37,295
H-411E	1		Post Synthesis Cooler #5/Mol Sieve preheater #2	0.04	$20,989	2002	$20,989	0.6	$2,991	2.47	$7,389	$8,745
H-412	1		Post Mixed Alcohol Cooler	0.61	$90,820	2002	$90,820	0.44	$73,284	2.47	$181,011	$214,230
H-413	1		Mixed Alcohol first Condenser (air cooled)	0.76	$42,255	1990	$42,255	1	$42,255	2.47	$104,369	$136,649
H-414	1		Mixed Alcohol Condenser	0.06	$338,016	2002	$338,016	0.44	$96,403	2.47	$238,116	$281,814
H-416B	1		Recycle Syngas Heat #2 / Flue gas Cool	0.63	$144,006	2002	$144,006	0.44	$117,700	2.47	$290,718	$344,070
K-410	1		Mixed Alcohol Gas Compressor	0.89	$851,523	2002	$851,523	0.8	$773,871	2.47	$1,911,462	$2,262,251
K-412	1		Purge Gas Expander	12.01	$642,014	2002	$642,014	0.8	$4,689,955	2.47	$11,584,188	$13,710,103
R-410	1		Mixed Alcohol Reactor	0.34	$2,026,515	2002	$2,026,515	0.56	$1,101,031	2.47	$2,719,545	$3,218,633
S-501	1		Mixed Alcohols Condensation Knock-out	2.01	$55,447	2002	$55,447	0.6	$84,229	2.47	$208,045	$246,225
A400						**Subtotal**	**$4,267,792**		**$7,046,667**	2.47	**$17,405,267**	**$20,612,585**
D-504	1		Ethanol/Propanol Splitter	0.50	$478,100	1998	$478,100	1.32	$189,541	2.1	$398,035	$478,460
D-505	1		Methanol/Ethanol Splitter	0.54	$478,100	1998	$478,100	1.32	$210,444	2.1	$441,933	$531,228
H-503A	1		Syngas Cooler #4 / Mol Sieve preheater #1	1.08	$20,989	2002	$20,989	0.6	$21,965	2.47	$54,255	$64,211
H-503B	1		Mol Sieve Superheater / reformed syngas cool #5	0.66	$4,743	2002	$4,743	0.6	$3,709	2.47	$9,161	$10,842
H-504C	1		D-504 condenser (air cooled)	0.68	$39,408	1990	$39,408	1	$39,408	2.47	$97,338	$127,442

(Continued)

Equipment Number	Number Required	Number Spares	Equipment Name	Size Ratio	Original Equip Cost (per unit)	Base Year	Total Original Equip Cost (Req'd & Spare) in Base Year	Scaling Exponent	Scaled Cost in Base Year	Installation Factor	Installed Cost in Base Year	Installed Cost in 2005$
H-504R	1		Ethanol/Propanol Splitter Reboiler	0.18	$158,374	1996	$158,374	0.68	$49,237	2.1	$103,398	$126,830
H-505C	1		D-505 condenser (air cooled)	0.64	$62,938	1990	$62,938	1	$62,938	2.47	$155,458	$203,538
H-505R	1		Methanol/Ethanol Splitter Reboiler	0.26	$158,374	1996	$158,374	0.68	$63,019	2.1	$132,340	$162,330
H-513	1		Mol Sieve Flush Condenser (air cooled)	0.19	$21,181	1990	$21,181	1	$21,181	2.47	$52,318	$68,499
H-590	1		MA Product Cooler / Mol Sieve preheater #3	0.34	$3,043	2002	$3,043	0.6	$1,590	2.47	$3,928	$4,649
H-591	2		Ethanol Product Pump	0.21	$7,501	1998	$15,002	1.79	$957	3.47	$3,321	$3,992
H-592	1		ETHANOL Product Cooler / Mol Sieve preheater #4	1.51	$3,043	2002	$3,043	0.6	$3,903	2.47	$9,639	$11,408
H-593	3		Ethanol Product Pump	1.19	$7,502	1999	$22,506	2.79	$36,692	4.47	$164,012	$196,596
P-504B	1	1	Ethanol/Propanol Splitter Bottoms Pump	0.04	$42,300	1997	$84,600	0.79	$7,120	2.8	$19,937	$24,151
P-504R	1	1	Ethanol/Propanol Splitter Reflux Pump	0.37	$1,357	1998	$2,714	0.79	$1,240	2.8	$3,471	$4,172
P-505B	1	1	Methanol/Ethanol Splitter Bottoms Pump	0.08	$42,300	1997	$84,600	0.79	$11,343	2.8	$31,761	$38,475
P-505R	1	1	Methanol/Ethanol Splitter Reflux Pump	0.37	$1,357	1998	$2,714	0.79	$1,240	2.8	$3,471	$4,172
P-590	1		Mixed Alcohol Product Pump	0.21	$7,500	1997	$7,500	0.79	$2,226	2.47	$5,499	$6,661
P-592	1		Ethanol Product Pump	1.19	$7,500	1997	$7,500	0.79	$8,614	2.47	$21,276	$25,773
S-502	1		LP Syngas Separator	0.49	$55,447	2002	$55,447	0.6	$36,305	2.47	$89,673	$106,130
S-503	1		Molecular Sieve (9 pieces)	1.43	$904,695	1998	$904,695	0.7	$1,160,495	2.47	$2,866,422	$3,445,594

(Continued)

Equipment Number	Number Required	Number Spares	Equipment Name	Size Ratio	Original Equip Cost (per unit)	Base Year	Total Original Equip Cost (Req'd & Spare) in Base Year	Scaling Exponent	Scaled Cost in Base Year	Installation Factor	Installed Cost in Base Year	Installed Cost in 2005$
T-504	1		Ethanol/Propanol Splitter Reflux Drum	0.37	$11,900	1997	$11,900	0.93	$4,731	2.1	$9,934	$12,034
T-505	1		Methanol/Ethanol Splitter Reflux Drum	0.37	$11,900	1997	$11,900	0.93	$4,731	2.1	$9,934	$12,034
T-590	2		Mixed Alcohol Product Storage Tank	0.21	$165,800	1997	$331,600	0.51	$151,390	2.47	$373,933	$452,976
T-592	2		Ethanol Product Storage Tank	1.19	$165,800	1997	$331,600	0.51	$362,599	2.47	$895,620	$1,084,940
A500						**Subtotal**	**$3,302,572**		**$2,456,617**	2.4244 99069	**$5,956,067**	**$7,207,140**
H-601	1		Steam Turbine Condenser	0.87	$0	2002	$0	0.71	$0	2.47	$0	$0
H-602	1		Blowdown Cooler / Deaerator Water Preheater	0.00	$3,043	2002	$3,043	0.6	$0	2.47	$0	$0
H-603	1		Blowdown Water-cooled Cooler	4.90	$16,143	2002	$16,143	0.44	$32,485	2.47	$80,237	$94,962
M-601	1		Hot Process Water Softener System	1.00	$1,031,023	1999	$1,031,023	0.82	$1,028,430	2.47	$2,540,222	$3,044,885
M-602	1		Extraction Steam Turbine/Generator	1.00	$4,045,870	2002	$4,045,870	0.71	$4,037,059	2.47	$9,971,535	$11,801,498
M-603	1		Startup Boiler	1.00	$198,351	2002	$198,351	0.6	$198,351	2.47	$489,927	$579,837
P-601	1	1	Collection Pump	0.13	$7,015	2002	$14,030	0.33	$7,226	2.47	$17,847	$21,122
P-602	1	1	Condensate Pump	0.87	$5,437	2002	$10,874	0.33	$10,366	2.47	$25,605	$30,304
P-603	1	1	Deaerator Feed Pump	1.00	$8,679	2002	$17,358	0.33	$17,340	2.47	$42,831	$50,691
P-604	1	1	Boiler Feed Water Pump	1.00	$95,660	2002	$191,320	0.33	$191,126	2.47	$472,082	$558,718
S-601	1		Blowdown Flash Drum	1.00	$14,977	2002	$14,977	0.6	$14,950	2.47	$36,926	$43,703
T-601	1		Condensate Collection Tank	1.00	$24,493	2002	$24,493	0.6	$24,448	2.47	$60,386	$71,468

(Continued)

Equipment Number	Number Required	Number Spares	Equipment Name	Size Ratio	Original Equip Cost (per unit)	Base Year	Total Original Equip Cost (Req'd & Spare) in Base Year	Scaling Exponent	Scaled Cost in Base Year	Installation Factor	Installed Cost in Base Year	Installed Cost in 2005$
T-602	1		Condensate Surge Drum	1.00	$28,572	2002	$28,572	0.6	$28,519	2.47	$70,443	$83,371
T-603	1		Deaerator	1.00	$130,721	2002	$130,721	0.72	$130,432	2.47	$322,168	$381,292
T-604	1		Steam Drum	1.00	$9,200	1997	$9,200	0.72	$9,180	2.47	$22,674	$27,467
A600						**Subtotal**	**$5,735,975**		**$5,729,912**	2.47	**$14,152,883**	**$16,789,318**
K-701	2	1	Plant Air Compressor	1.00	$32,376	2002	$97,129	0.34	$97,129	2.47	$239,908	$283,936
M-701	1		Cooling Tower System	0.66	$267,316	2002	$267,316	0.78	$193,008	2.47	$476,730	$564,218
M-702	1		Hydraulic Truck Dump with Scale	1.00	$80,000	1998	$80,000	0.6	$80,000	2.47	$197,600	$237,526
M-703	1		Flue Gas Stack	0.31	$51,581	2002	$51,581	1	$15,917	2.47	$39,315	$46,530
P-701	1	1	Cooling Water Pump	0.66	$158,540	2002	$317,080	0.33	$276,264	2.47	$682,373	$807,601
P-702	1	1	Firewater Pump	1.00	$18,400	1997	$36,800	0.79	$36,800	2.47	$90,896	$110,110
P-703	1	1	Diesel Pump	1.00	$6,100	1997	$12,200	0.79	$12,200	2.47	$30,134	$36,504
P-704	1	1	Ammonia Pump	1.00	$5,000	1997	$10,000	0.79	$10,000	2.47	$24,700	$29,921
P-705	1		Hydrazine Pump	1.00	$5,500	1997	$5,500	0.79	$5,500	2.47	$13,585	$16,457
S-701	1	1	Instrument Air Dryer	1.00	$8,349	2002	$16,698	0.6	$16,698	2.47	$41,244	$48,813
T-701	1		Plant Air Receiver	1.00	$7,003	2002	$7,003	0.72	$7,003	2.47	$17,297	$20,472
T-702	1		Firewater Storage Tank	1.00	$166,100	1997	$166,100	0.51	$166,100	2.47	$410,267	$496,991
T-703	1		Diesel Storage Tank	1.00	$14,400	1997	$14,400	0.51	$14,400	2.47	$35,568	$43,086
T-704	1		Ammonia Storage Tank	1.00	$287,300	1997	$287,300	0.72	$287,300	2.47	$709,631	$859,635
T-705	1		Olivine Lock Hopper	1.00	$0	1998	$0	0.71	$0	2.47	$0	$0
T-706	1		MgO Lock Hopper	1.00	$0	1998	$0	0.71	$0	2.47	$0	$0
T-707	1		Hydrazine Storage Tank	1.00	$12,400	1997	$12,400	0.93	$12,400	2.47	$30,628	$37,102
A700						**Subtotal**	**$1,381,507**		**$1,230,719**	2.47	**$3,039,875**	**$3,601,800**
					Equipment Cost		**$40,920,110**		**$46,915,218**	2.4676 17439	**$115,768,810**	**$137,228, 869**

Appendix F. Discounted Cash Flow Rate of Return Summary

DCFROR Worksheet													
Year		-2	-1	0	1	2	3	4	5	6	7	8	9
Fixed Capital Investment		$18,603,984	$114,503,967	$61,068,782									
Working Capital				$9,541,997									
Loan Payment					$0	$0	$0	$0	$0	$0	$0	$0	$0
Loan Interest Payment		$0	$0	$0	$0	$0	$0	$0	$0	$0	$0	$0	$0
Loan Principal		$0	$0	$0	$0	$0	$0	$0	$0	$0	$0	$0	$0
Ethanol Sales					$46,938,097	$62,584,129	$62,584,129	$62,584,129	$62,584,129	$62,584,129	$62,584,129	$62,584,129	$62,584,129
By-Product Credit					$9,633,280	$12,844,373	$12,844,373	$12,844,373	$12,844,373	$12,844,373	$12,844,373	$12,844,373	$12,844,373
Total Annual Sales					$56,571,377	$75,428,503	$75,428,503	$75,428,503	$75,428,503	$75,428,503	$75,428,503	$75,428,503	$75,428,503
Annual Manufacturing Cost													
Raw Materials					$23,647,650	$27,025,885	$27,025,885	$27,025,885	$27,025,885	$27,025,885	$27,025,885	$27,025,885	$27,025,885
Tar reforming catalysts					$808,613								
Steam reforming catalysts					$0					$0			
ZnO					$0					$0			
Mixed Alcohol catalysts					$542,966	$0	$0	$0	$0	$542,966	$0	$0	$0
Baghouse Bags					$415,430					$415,430			
Other Variable Costs					$1,531,320	$1,732,166	$1,732,166	$1,732,166	$1,732,166	$1,732,166	$1,732,166	$1,732,166	$1,732,166
Fixed Operating Costs					$12,059,682	$12,059,682	$12,059,682	$12,059,682	$12,059,682	$12,059,682	$12,059,682	$12,059,682	$12,059,682
Total Product Cost					$39,005,661	$40,817,733	$40,817,733	$40,817,733	$40,817,733	$41,776,129	$40,817,733	$40,817,733	$40,817,733
Annual Depreciation													
General Plant													
DDB					$38,064,676	$27,189,054	$19,420,753	$13,871,967	$9,908,548	$7,077,534	$5,055,381		
SL					$19,032,338	$15,860,282	$13,594,527	$12,137,971	$11,559,972	$11,559,972	$11,559,972		
Remaining Value					$95,161,690	$67,972,636	$48,551,883	$34,679,916	$24,771,369	$17,693,835	$12,638,453		
Actual					$38,064,676	$27,189,054	$19,420,753	$13,871,967	$11,559,972	$11,559,972	$11,559,972		
Steam Plant													
DDB					$4,321,018	$3,996,942	$3,697,171	$3,419,883	$3,163,392	$2,926,138	$2,706,677	$2,503,677	$2,315,901
SL					$2,880,679	$2,804,872	$2,738,645	$2,682,262	$2,636,160	$2,601,011	$2,577,788	$2,567,873	$2,567,873
Remaining Value					$53,292,560	$49,295,618	$45,598,446	$42,178,563	$39,015,171	$36,089,033	$33,382,355	$30,878,679	$28,562,778
Actual					$4,321,018	$3,996,942	$3,697,171	$3,419,883	$3,163,392	$2,926,138	$2,706,677	$2,567,873	$2,567,873
Net Revenue					($24,819,978)	$3,424,773	$11,492,845	$17,318,919	$19,887,405	$19,166,264	$20,344,120	$32,042,896	$32,042,896
Losses Forward						($24,819,978)	($21,395,205)	($9,902,360)	$0	$0	$0	$0	$0
Taxable Income					($24,819,978)	($21,395,205)	($9,902,360)	$7,416,559	$19,887,405	$19,166,264	$20,344,120	$32,042,896	$32,042,896
Income Tax					$0	$0	$0	$2,892,458	$7,756,088	$7,474,843	$7,934,207	$12,496,729	$12,496,729
Annual Cash Income					$17,565,716	$34,610,769	$34,610,769	$31,718,311	$26,854,681	$26,177,531	$26,676,563	$22,114,040	$22,114,040
Discount Factor		1.21	1.1	1	0.909090909	0.826446281	0.751314801	0.683013455	0.620921323	0.56447393	0.513158118	0.46650738	0.424097618
Annual Present Value	$217,657,607				$15,968,833	$28,603,942	$26,003,583	$21,664,033	$16,674,644	$14,776,534	$13,689,295	$10,316,363	$9,378,512
Total Capital Investment + Interest		$22,510,821.23	$125,954,363.41	$70,610,779.49									
Net Present Worth				$0									

(Continued)

DCFROR Worksheet											
Year	10	11	12	13	14	15	16	17	18	19	20
Fixed Capital Investment											
Working Capital											($9,541,997)
Loan Payment	$0	$0	$0	$0	$0	$0	$0	$0	$0	$0	$0
Loan Interest Payment	$0	$0	$0	$0	$0	$0	$0	$0	$0	$0	$0
Loan Principal	$0	$0	$0	$0	$0	$0	$0	$0	$0	$0	$0
Ethanol Sales	$62,584,129	$62,584,129	$62,584,129	$62,584,129	$62,584,129	$62,584,129	$62,584,129	$62,584,129	$62,584,129	$62,584,129	$62,584,129
By-Product Credit	$12,844,373	$12,844,373	$12,844,373	$12,844,373	$12,844,373	$12,844,373	$12,844,373	$12,844,373	$12,844,373	$12,844,373	$12,844,373
Total Annual Sales	$75,428,503	$75,428,503	$75,428,503	$75,428,503	$75,428,503	$75,428,503	$75,428,503	$75,428,503	$75,428,503	$75,428,503	$75,428,503
Annual Manufacturing Cost											
Raw Materials	$27,025,885	$27,025,885	$27,025,885	$27,025,885	$27,025,885	$27,025,885	$27,025,885	$27,025,885	$27,025,885	$27,025,885	$27,025,885
Tar reforming catalysts											
Steam reforming catalysts		$0					$0				
ZnO		$0					$0				
Mixed Alcohol catalysts	$0	$542,966	$0	$0	$0	$0	$542,966	$0	$0	$0	$0
Baghouse Bags		$415,430					$415,430				
Other Variable Costs	$1,732,166	$1,732,166	$1,732,166	$1,732,166	$1,732,166	$1,732,166	$1,732,166	$1,732,166	$1,732,166	$1,732,166	$1,732,166
Fixed Operating Costs	$12,059,682	$12,059,682	$12,059,682	$12,059,682	$12,059,682	$12,059,682	$12,059,682	$12,059,682	$12,059,682	$12,059,682	$12,059,682
Total Product Cost	$40,817,733	$41,776,129	$40,817,733	$40,817,733	$40,817,733	$40,817,733	$41,776,129	$40,817,733	$40,817,733	$40,817,733	$40,817,733
Annual Depreciation											
General Plant											
DDB											
SL											
Remaining Value											
Actual											
Steam Plant											
DDB	$2,142,208	$1,981,543	$1,832,927	$1,695,457	$1,568,298	$1,450,676	$1,341,875	$1,241,234	$1,148,142	$1,062,031	$982,379
SL	$2,567,873	$2,567,873	$2,567,873	$2,567,873	$2,567,873	$2,567,873	$2,567,873	$2,567,873	$2,567,873	$2,567,873	$2,567,873
Remaining Value	$26,420,570	$24,439,027	$22,606,100	$20,910,642	$19,342,344	$17,891,668	$16,549,793	$15,308,559	$14,160,417	$13,098,386	$12,116,007
Actual	$2,567,873	$2,567,873	$2,567,873	$2,567,873	$2,567,873	$2,567,873	$2,567,873	$2,567,873	$2,567,873	$2,567,873	$2,567,873
Net Revenue	$32,042,896	$31,084,500	$32,042,896	$32,042,896	$32,042,896	$32,042,896	$31,084,500	$32,042,896	$32,042,896	$32,042,896	$32,042,896
Losses Forward	$0	$0	$0	$0	$0	$0	$0	$0	$0	$0	$0
Taxable Income	$32,042,896	$31,084,500	$32,042,896	$32,042,896	$32,042,896	$32,042,896	$31,084,500	$32,042,896	$32,042,896	$32,042,896	$32,042,896
Income Tax	$12,496,729	$12,122,955	$12,496,729	$12,496,729	$12,496,729	$12,496,729	$12,122,955	$12,496,729	$12,496,729	$12,496,729	$12,496,729
Annual Cash Income	$22,114,040	$21,529,419	$22,114,040	$22,114,040	$22,114,040	$22,114,040	$21,529,419	$22,114,040	$22,114,040	$22,114,040	$22,114,040
Discount Factor	0.385543289	0.350493899	0.318630818	0.28966438	0.263331254	0.239392049	0.217629136	0.197844669	0.17985879	0.163507991	0.148643628
Annual Present Value	$8,525,920	$7,545,930	$7,046,215	$6,405,650	$5,823,318	$5,293,925	$4,685,429	$4,375,145	$3,977,404	$3,615,822	$3,287,111
Total Capital Investment + Interest											($1,418,357.09)
Net Present Worth											

Ethanol from Mixed Alcohols Production Process Engineering Analysis

2012 Market Target Case
2,000 Dry Metric Tonnes Biomass per Day
BCL Gasifier, Tar Reformer, Sulfur Removal, MoS2 Catalyst, Fuel Purification, Steam-Power cycle
All Values in 2005$

Minimum Ethanol Selling Price ($/gal) **$1.01**

EtOH Production at Operating Capacity (MM Gal / year) 61.8
EtOH Product Yield (gal / Dry US Ton Feedstock) 80.1
Mixed Alcohols Production at Operating Capacity (MM Gal / year) 72.6
Mixed Alcohols Product Yield (gal / Dry US Ton Feedstock) 94.1
Delivered Feedstock Cost $/Dry US Ton $35 Internal Rate of Return (After-Tax) 10%
Equity Percent of Total Investment 100%

Capital Costs		**Operating Costs (cents/gal product)**	
Feed Handling & Drying	$23,200,000	Feedstock	43.7
Gasification	$12,900,000	Natural Gas	0.0
Tar Reforming & Quench	$38,400,000	Catalysts	0.3
Acid Gas & Sulfur Removal	$14,500,000	Olivine	0.7
Alcohol Synthesis - Compression	$16,000,000	Other Raw Materials	1.6
Alcohol Synthesis - Other	$4,600,000	Waste Disposal	0.4
Alcohol Separation	$7,200,000	Electricity	0.0
Steam System & Power Generation	$16,800,000	Fixed Costs	19.5
Cooling Water & Other Utilities	$3,600,000	Co-product credits	-20.7
Total Installed Equipment Cost	$137,200,000	Capital Depreciation	15.4
		Average Income Tax	11.8
Indirect Costs	53,600,000	Average Return on Investment	28.5
(% of TPI)	28.1%		
Project Contingency	4,100,000	**Operating Costs ($/yr)**	
		Feedstock	$27,000,000
Total Project Investment (TPI)	$190,800,000	Natural Gas	$0
		Catalysts	$200,000
Installed Equipment Cost per	$2.22	Olivine	$400,000
Total Project Investment per Annual	$3.09	Other Raw Matl. Costs	$300,000
		Waste Disposal	$300,000
Loan Rate	N/A	Electricity	$0
Term (years)	N/A	Fixed Costs	$12,100,000
Capital Charge Factor	0.180	Co-product credits @ $1.15 per gal	-
		Capital Depreciation	$9,500,000
Maximum Yields based on carbon		Average Income Tax	$7,300,000
Theoretical Ethanol Production	158.9	Average Return on Investment	$17,600,000
Theoretical Ethanol Yield (gal/dry	205.8		
Current Ethanol Yield	39%	Total Plant Electricity Usage (KW)	7,994
		Electricity Produced Onsite (KW)	7,998
Gasifier Efficiency - HHV %	76.6	Electricity Purchased from Grid	0
Gasifier Efficiency - LHV %	76.1	Electricity Sold to Grid (KW)	4
Overall Plant Efficiency - HHV %	47.4		
Overall Plant Efficiency - LHV %	45.8	Steam Plant + Turboexpander Power	66,451
		Used for Main Compressors (hp)	55,168

(Continued)

Capital Costs		Operating Costs (cents/gal product)	
Plant Hours per year	8406	Used for Electricity Generation (hp)	11,283
%	96.0%	Plant Electricity Use (KWh/gal product)	1.5
		Gasification & Reforming Steam Use (lb/gal)	9.9

APPENDIX G. PROCESS PARAMETERS & OPERATION SUMMARY

Energy Efficiencies		Tar Reformer		Alcohol Synthesis		Alcohol Synthesis	
Gasifier Efficiency – HHV %	76.6	Inlet Molar Flow (MMscf/hr)	6.70	Syngas from Conditioning	279,888	Relative Alcohol Distribution After Reactor	
Gasifier Efficiency – LHV %	76.1	Space Velocity (hr^{-1})	2,476	Recycled from initial flashtank	215	Methanol	8.5%
Overall Plant Efficiency - HHV%	47.4	Reactor Volume (ft^3)	2,705	Recycled from MolSieve Flush	5,026	Ethanol	81.7%
Overall Plant Efficiency - -LHV %	45.8	Inlet:		Total	285,128	Propanol	8.8%
		Temperature (°F)	1,086			Butanol	0.9%
Dryer		Carbon as CO (mol%)	49.0%			Pentanol +	0.1%
Inlet:		Carbon as tar (ppmv)	5,758	Conditioned Syngas H_2:CO Ratio	1.00		
Temperature (°F)	60.0	H_2:CO Ratio (mole)	0.87	Recycled Gas H_2:CO Ratio	1.06	Flash Separator	
Moisture Content (wt%)	50.0%					Temp-erature (°F)	110
Outlet:		Reformer Conversions:		At Reactor Inlet		Pressure (psia)	970
Temperature (°F)	219.7	CO_2 --> CO	32.5%	Temperature (°F)	570		
Moisture Content (wt%)	5.0%	Methane --> CO	80.0%	Pressure (psia)	991	Relative Alcohol Distribution After Flash Tank	
Inlet Flue Gas (°F)	1,206	Ethane --> CO	99.0%	H2:CO Molar Ratio	1.00	Methanol	8.3%
Outlet Flue Gas (°F)	235.9	Ethylene -- > CO	90.0%	CO_2 (mol %)	5.0%	Ethanol	81.7%
Dew Point Flue Gas (°F)	173.9	Benzene --> CO	99.0%	Methane (mol%)	1.5%	Propanol	8.9%

(Continued)

Energy Efficiencies		Tar Reformer		Alcohol Synthesis		Alcohol Synthesis	
Difference	62.0	Tar --> CO	99.9%	H2O (wt%)	0.64%	Butanol	0.9%
		Ammonia --> CO	90.0%			Pentanol +	0.1%
Gasifier				Inlet Molar Flow (MMscf/hr)	6.5		
Temperature (°F)	1,633	Outlet:		Space Velocity (hr^{-1})	4,000	Vapor Losses From Flash Tank	
Pressure (psia)	23.0	Temperature (°F)	1,633	Reactor Volume (ft^3)	1,616	Methanol	3.5%
H2:CO Molar Ratio After Gasifier	0.60	Carbon as CO (mol%)	75.7%			Ethanol	1.8%
Methane (vol%)	9.0%	Carbon as tar (ppmv)	43	CO Conversion - Overall	59.5%	Propanol	0.8%
Benzene (vol%)	0.07%	H_2:CO Ratio (mole)	1.00	CO Conversion - Singlepass	59.4%	Butanol	0.3%
Tar (wt%)	0.91%	Methane (vol%)	1.2%	Conversion To:		Pentanol +	0.1%
Tar (g/Nm^3)	9.5	Benzene (ppmv)	2.7	CO_2	21.9%		
Char (wt%)	12.7%	Tars (ppmv)	3	Methane	3.4%	Cleaned Gas Recyc-led toReac-tor	0.1%
H2S (ppm)	413	Tars (g/Nm^3)	0.01	Ethane	0.3%		
Residual Heat (MBtu/hr)	0	H_2S (ppm)	205	Methanol	0.3%	Residual Syngas	
		NH3 (ppm)	80	Ethanol	28.2%	Recycled to synthesis reactors (lb/hr)	215
Char Combustor				Propanol	4.6%	To Tar Reormers (lb/hr)	214,787
Temperature (°F)	1,823	***Quench***		Butanol	0.6%	To Fuel Sysem (lb/hr)	10,739
Pressure (psia)	21.4	Benzene (ppmv)	3.1	Pentanol +	0.1%	To Reform-er for Proc-ess (lb/hr)	204,048
Ratio Actual: Minimum air for combustion	1.20	Tars (ppmv)	4	Total	59.4%		
Residual Heat (MBtu/hr)	0.0	Tars (g/Nm^3)	0.01			Overall Water Demand	
		H_2S (ppm)	235	Selectivity (CO_2 Free)		gal/gal etoh	1.94
Syngas Usage		NH3 (ppm)	83	Alcohols	90.1%	gal/gal total alcohols	1.65

(Continued)

Energy Efficiencies		Tar Reformer		Alcohol Synthesis		Alcohol Synthesis	
To Reformer (lb/hr)	168,120			Hydrocarbons	9.9%		
To Char Combustor (lb/hr)	0	***Acid Gas Removal***					
To Fuel System (lb/hr)	64,703	Inlet:		At Reactor Outlet			
		CO_2 (mol/hr)	2,041	Temperature (°F)	570		
Fuel System		CO_2 (mol%)	11.4%	Pressure (psia)	986		
Additional fuel (lb/hr)	0	H_2S (mol/hr)	4	CO_2 (mol%)	22.7%		
Raw Syngas (lb/hr)	64,703	H_2S (ppmv)	235	Methane (mol%)	4.6%		
Unconverted Syngas (lb/hr)	10,739	Outlet:		H_2O (wt%)	0.47%		
		CO_2 (mol/hr)	846				
Into Reformer (°F)	3,680	CO_2 (mol%)	5.1%	Total Alcohol Productivity (kg/kg/hr)	0.602		
Out of Reformer (°F)	1,780	Fraction CO_2 removed	58.6%	Total Ethanol Productivity (kg/kg/hr)	0.488		
		H_2S (mol/hr)	1				
		H_2S (ppmv)	49				
		Fraction H_2S removed	99.6%				

APPENDIX H. SYNGAS AND CHAR CORRELATIONS

The gasifier was modeled using correlations based on data from the Battelle Columbus Laboratory (BCL) 9 tonne/day test facility. The data and original correlations for the gasifier can be found in Bain (1992). The experimental runs were performed for several different wood types including Red Oak, Birch, Maple, and Pine chips, sawdust, and other hard and soft wood chips. The original pilot plant data for these runs can be found in Feldmann, et al, (1988). The temperature range for the data is 1,280 to 1,857°F and the pressure range is 2.4 to 14.4 psig; the majority of the data are in the range of 1,500 to 1,672°F.

The BCL test facility's gas production data was correlated to gasifier temperature with a quadratic function in the form:

$$X = a + bT + cT^2$$

where the temperature, T, in units of °F. The coefficients a, b, and c, as well as the units for the correlated variable are shown in Table 3. Even thought there is a correlation for the char formation, it is not used; instead the amount and elemental analysis for the char is determined by mass differences between the produced syngas and the converted biomass.

The following general procedure is used for the gasifier production:

- A gasifier temperature T is assumed.

- The mass and molar amounts of carbon, hydrogen, oxygen, sulfur, nitrogen, and ash (as a pseudo-element) are determined from the biomass's ultimate analysis.
- The amount of syngas and its composition is determined from the gasifier correlations.
- The amount of carbon in the syngas and tar is determined. Residual carbon is parsed in the char.
- The amount of oxygen in the syngas is determined. A minimum amount of oxygen is required to be parsed to the char (4% of biomass oxygen). If there is a deficit of oxygen, then the associated water is decomposed to make sure that this amount of oxygen is parsed to the char; if there is excess oxygen, then it is parsed to the char without decomposing hydrogen.
- A set amount of sulfur is parsed to the char (8.3%). All remaining sulfur is set as H_2S in the syngas.
- A set amount of nitrogen is parsed to the char (6.6%). All remaining nitrogen is set as NH_3 in the syngas.
- The amount of hydrogen in the syngas (including tar, H_2S, NH_3, and decomposed water) is determined. All remaining hydrogen is parsed to the char.
- All ash is parsed to the char.
- The heat of formation of the char is estimated from the resulting ultimate analysis from this elemental material balance.
- The gasifier temperature is adjusted so the there is no net heat for an adiabatic reaction.

The syngas amount and composition will be dependent upon the biomass composition and the gasifier temperature. As an example, the resulting syngas composition for the woody biomass used in this design report can be seen in Figure 2. Note from this figure that the amount of char decreases with increasing temperature and that the water does not start to decompose until high temperatures (here at 1650°F and higher).

REFERENCES

Bain, R; (January 14, 1992). *Material and Energy Balances for Methanol from Biomass Using Biomass Gasifiers.*

Feldmann, HF; Paisley, MA; Applebaus, HR; Taylor, DR. (July 1988). *Conversion of Forest Residues to A Medium-Rich Gas in a High-Throughput Gasifier.*

Table 3. Gasifier Correlation

Variable	*a*	*b*	*c*	Units
Dry Syngas	28.993	-0.043325	0.000020966	scf gas/lb maf wood[a]
CO	133.46	-0.1029	0.000028792	mol% dry gas
CO_2	-9.525 1	0.037889	-0.000014927	mol% dry gas
CH_4	-13.82	0.044179	-0.000016167	mol% dry gas
C_2H_4	-38.258	0.058435	-0.000019868	mol% dry gas

Table 3 (Continued)

C_2H_6	11.114	-0.011667	0.000003064	mol% dry gas
H_2	17.996	-0.026448	0.00001893	mol% dry gas
C_2H_2	-4.3114	0.0054499	-0.000001561	mol% dry gas
Tar	0.045494	-0.0000 19759		lb/lb dry wood

[a] Scf = standard cubic feet. The standard conditons are 1 atm pressure and 60°F temperature.

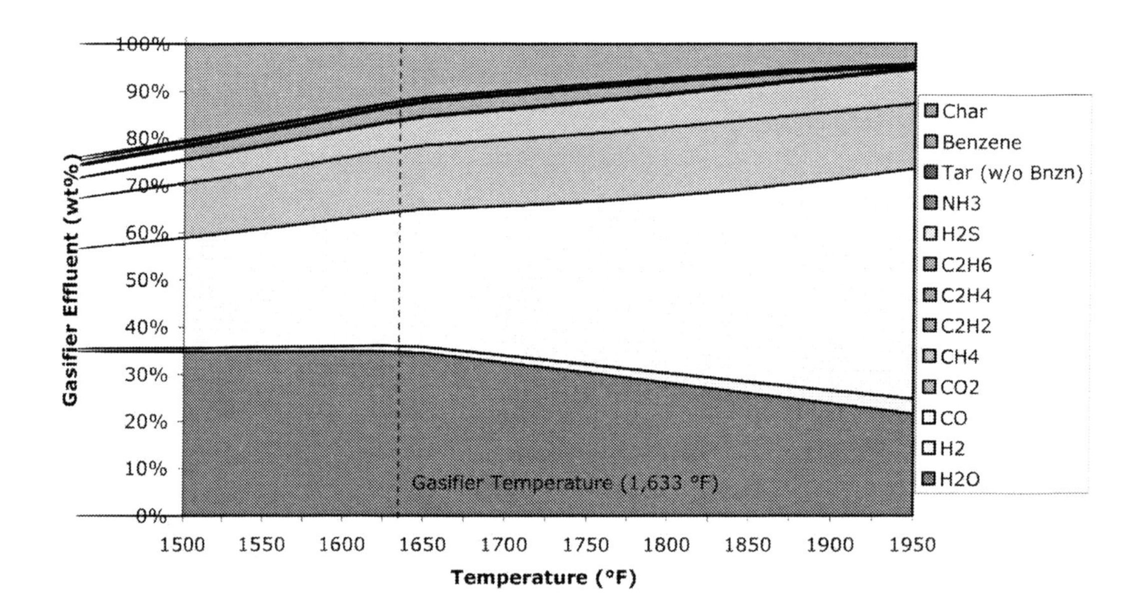

Figure 2. Syngas Composition for Woody Biomass Used in Design Report

APPENDIX I. ALCOHOL SYNTHESIS CATALYST REFERENCES

A literature search was conducted to review existing mixed alcohol technology and how it has developed over the past 20 years. This appendix provides a list of literature reviewed, including authors and journal names, as well as brief descriptions of the information contained within each source. The literature sources examined include a mixture of experimental results and state of technology summarizations. Together, over 40 different sources were used. Some generalizations garnered from this search are summarized below.

The term "mixed alcohols" refers to a mixture of C1 – C8 alcohols, with preference towards the higher alcohols (C2-C6). Mixed alcohols catalysts are typically categorized into several groups based on their composition and/or derivation. Common to all of these catalysts is the addition of alkali metals which shift the product slate towards alcohol production. Spath, et al., categorized the catalysts into five groupings based on the work of Herman (Herman, 1991):

- Modified high pressure methanol catalysts.
- Modified low pressure methanol catalysts.
- Modified Fischer-Tropsch (FT) catalysts.

- Alkali-doped sulfides (modified methanation).
- Other, which includes alternate catalysts, such as Rhodium based catalysts, which are not specifically used for mixed alcohols but have been developed for more selective alcohols synthesis.

Others (Smith, et al., 1992) (Forzatti, et al., 1991) group them simply into three categories:

- Modified methanol catalysts.
- Modified Fischer-Tropsch catalysts.
- Others.

This helps to eliminate confusion that can arise when, for example, molybdenum-based sulfide catalysts are promoted with cobalt or other similar FT elements, thus representing both alkali- doped sulfides and modified-FT groupings.

Since the 1 920s scientists have known how to produce mixtures of methanol and other alcohols by reacting syngas over certain catalysts. They observed that when methanol catalysts (Zinc or Copper based) were promoted with alkali (Na, K, etc.), and certain reaction conditions were met (temperature, pressure) a mixture of methanol and higher alcohols resulted. At the same time, Fischer and Tropsch observed that hydrocarbon synthesis catalysts produced linear alcohols as byproducts. From this they were able to develop the "Synthol" process for producing higher alcohols. Some development continued, but it wasn't until the 1970s oil embargo that significant interest re-appeared, and researchers renewed efforts to produce higher alcohols for liquid fuels applications. As petroleum prices dropped research declined until the mid-to-late 1980s when interest was driven by environmental aspects, specifically oxygenated fuel and octane enhancement.

In 1990, the Clean Air Act mandated the seasonal use of oxygenated compounds in gasoline in specific regions of the U.S. Soon after, methyl-tertiary-butyl ether (MTBE) became the oxygenate of choice because refiners could cost-effectively produce it using existing products. Since then only a few researchers have been active in the field of higher alcohol synthesis. Some research in the 90s focused on mixed alcohols as a product of coal gasification. Other work continued in Europe, especially by Snamprogetti. Within the past 5 years, however, a desire to find alternatives for petroleum based fuels, and the increasing popularity of ethanol fuels, has again brought this research area to life.

Three particular terms require definition before the discussion proceeds. Please refer back to these as needed. Within literature, data can often be confusing if these terms aren't well understood or defined.

Productivity – the amount of product generated, either all alcohols or a specific alcohol, per unit time for a certain weight of catalyst loaded. Units are typically g-product / kg catalyst / hr. Typical productivities for mixed alcohol production can range from below 150 to near 400 g/kg/hr. In comparison, methanol synthesis productivities can often be over 1000 g/kg/hr, or 1 kg/kg/hr. Less commonly, productivity refers to concentrations of liquid product per unit of time, g/L/hr. This is often referred to as "Yield".

Conversion – usually the amount of carbon monoxide (CO, molar basis) converted to all products. Typically found by: $(CO_{initial} - CO_{final}) / CO_{initial}$

Sometimes researchers present conversions exclusive of CO_2 produced but not often. Typical single-pass conversions for mixed alcohols range from 10% - 40%.

Selectivity – selectivities are typically presented on a %-molar basis. Selectivity refers to the fraction of CO converted to a specific product. For example, if a reaction achieves a 20% conversion of CO, and 75% of that CO (or 15% of the total CO) is converted to alcohols, then the total alcohols selectivity is said to be 75%. Selectivities are sometimes shown on a CO_2-exclusive basis.

Yield – see Productivity.

The overall stoichiometric reaction for higher alcohol synthesis (HAS) can be summarized as:

$$n\,CO + 2n\,H_2 \rightarrow C_nH_{2n+1}OH + (n-1)\,H_2O$$

The value of "*n*" typically ranges from 1 to 8. The stoichiometry suggests an optimum H2/CO ratio of 2, however many of these catalysts also display significant water-gas shift activity. This shifts the optimal ratio closer to 1.0 and also shifts the primary byproduct from water to carbon dioxide (CO_2). The overall reaction is exothermic; therefore, maintaining constant reaction temperature is an important design consideration. The reactions become more exothermic for greater values of "*n*". Secondary reactions and other side products will depend on which catalyst system is used. Different kinetic pathways exist for each catalyst system.

Table 4. "Typical" Modified Methanol Catalyst Conditions

	H_2/CO ratio	**Temperature (°F)**	**Pressure (psia)**	**CO conversion (per pass)**	**Total Alcohol Yield (g/kg/hr)**	**C_2+OH Selec-tivity**
High Pressure	1	572 to 800	1,810 to 3,625	5 to 20%	203	
Low Pressure*	1.0 to 1.2	482 to 752	725 to 1,450	20 to 60%		41.9 wt%

* Lurgi: Octamix

Catalysts

Modified Methanol Catalysts

The term "modified" methanol catalyst refers to the addition of an alkali promoter and other active elements to a methanol catalyst to shift the product slate from methanol to higher branched primary alcohols. High temperature methanol catalysts typically contain Zinc (Zn)

Chromium oxides (or manganese chromium oxides), while lower temperature methanol catalysts use Copper (Cu) as the active component. The reaction yields primary branched alcohols, among which 2-methyl-1 -propanol (isobutanol) is a main (and thermodynamically favored) component. Aldehydes, esters, ketones, and ethers are also formed, along with large amounts of CO_2.

Typical high pressure and low pressure reaction conditions (as provided by Nexant) are listed in Table 4. In general there is a trade-off between maximizing CO conversion and maximizing the higher alcohol selectivity and yield.

Snamprogetti (also referred to as SEHT – Snamprogetti, Enichem and Haldor Topsoe) and Lurgi were two of the leading technology developers of modified methanol catalysis in the 1980s and 1 990s. SEHT had a MAS (Metanolo piu Alcoli Superiori - methanol plus higher alcohols) process and Lurgi developed what they called OCTAMIX, each developing pilot scale plants and data. The latest information available to NREL shows each process technology is no longer available. One technology developer still involved in this area is the Standard Alcohol Company of America. They have a bench-scale process to produce a mixed alcohols product known as Envirolene^{TM}. Envirolene is composed of methanol through octanol, with approximately 50% of the product as ethanol. The process uses a modified high pressure methanol catalyst, and the company is currently seeking funding for a pilot plant.

The proposed kinetic pathway for modified methanol catalysts to branched alcohols is through a base-catalyzed aldol condensation reaction. Carbon chain growth schemes have been developed that describe the product distribution relatively accurately. This is shown in Figure 3.

Modified Fischer-Tropsch Catalysts

Modified Fischer-Tropsch catalysts, on the other hand, are FT catalysts that are alkali-promoted. The two most common FT active elements are Iron (Fe) and Cobalt (Co), but Nickel (Ni) is considered to have FT activity also. The addition of the alkali promoter helps to shift the product slate from hydrocarbons to linear alcohols, although hydrocarbons remain a significant byproduct. Typical reaction conditions are 220 – 350°C (430 – 660°F) and 5-20 MPa (725 – 2,900 psia). One commonly-researched catalyst system is a MoS_2-based system that is alkali and/or Cobalt-promoted. This has the tendency to increase ethanol and other higher alcohols selectivity. CO_2 is still a substantial byproduct due to water-gas shift (WGS) activity of the catalysts. Other potential byproducts include aldehydes, esters, carboxylic acids, and ketones.

The primary technology developers for these catalysts were Dow/Union Carbide (UCC) and Institut Francais du Petrole (IFP). Dow and UCC jointly developed a sulfided mixed alcohol catalyst based on molybdenum (MoS_2). Sulfided catalysts have the advantage of being sulfur- tolerant (up to 100 ppm) which has the potential to reduce upstream cleanup costs. IFP, in conjunction with Idemitsu Kosan (Japan), developed a process based on Cu-Co and Cu-Ni catalyst systems. Dow built a 2-ton-per-day (TPD) demonstration plant in 1990 and IFP built a 20 barrel-per-day (BPD) pilot plant in Japan. The latest information available to NREL indicates that Dow is no longer pursuing the commercial development of their mixed alcohol process and IFP has not continued their work since building the pilot plant, and they have no commercial interest in pursuing a mixed alcohols process.

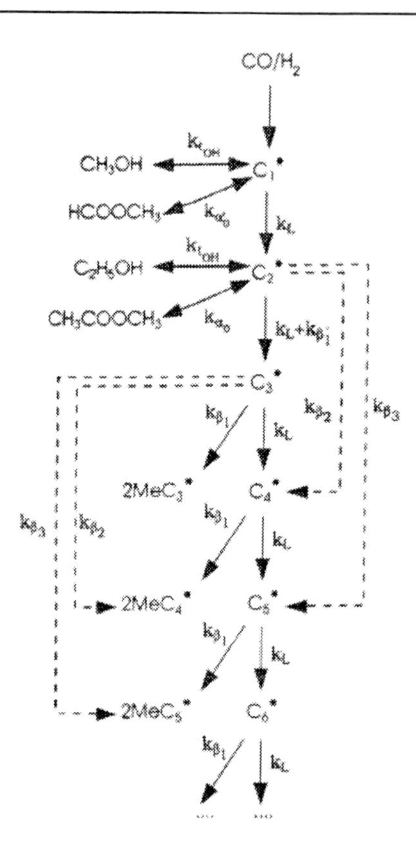

Figure 3. Reaction Network of Smith, et.al., 1991 for the Methanol – Higher Alcohol Synthesis over Cu/ZnO-based

Technology developers that remain active in this area are Power Energy Fuels Inc. (PEFI), Western Research Institute (WRI), and Pearson Technologies. PEFI continues to develop the Ecalene^{TM} technology and process, which is a modification of Dow's Sygmal process using polysulfite catalyst. According to Nexant, progress has not moved beyond the bench scale and a planned 500 gallon/day pilot plant is no longer being pursued. However, 2-3 other pilot plants are under funding consideration using various biomass resources. WRI had worked with PEFI in the past, however currently they are not, but are conducting their own bench-scale experiments, particularly reactor and catalyst testing. Pearson Technologies has developed a 30-ton-per-day biomass gasification and alcohols conversion facility in Aberdeen, MS. A project is under development by the Worldwide Energy Group and the State of Hawaii to demonstrate gasification of sugarcane bagasse and production of ethanol using the Pearson technology on the island of Kauai. Sasol (South Africa) is a world leader in FT fuels and chemicals production as well as technology development. They currently produce a mixture of alcohols within their overall process. However, these are not used for fuels. According to Sasol's website, oxygenates in the aqueous stream from their Sasol Advanced Synthol (SAS) process are separated and purified to produce alcohols, acetic acid, and ketones.

Due to the severe process conditions of higher alcohol synthesis, it would be expected that catalyst life would not be significantly long[b] (Nexant, 2005). As a benchmark, it could be helpful to recognize that the catalyst life for the typical Fe-Co Fischer-Tropsch catalyst can be longer than 5 years. Information on catalyst life is not abundant because most research into HAS is bench scale and not commercial. The majority of information on catalyst life comes from the earlier commercialization attempts. Dow and UCC, for example, found their catalyst operated for over 8 months continuously with little to no process performance degradation.

The proposed kinetics for modified FT catalysts follow different pathways than for modified methanol kinetics. Linear alcohols are formed from a classic CO insertion route for chain growth (C-C bond formation) with termination to alcohols and hydrocarbons. This is shown in Figure 4. More complex kinetic models have reaction networks that account for the simultaneous formation of alcohols, hydrocarbons, and esters.

Other Catalysts

Some research has been conducted on alternative reaction systems for mixed alcohols synthesis. This includes more exotic catalytic elements (Ruthenium (Ru), Rhodium (Rh), Palladium (Pd)) as well as synthesis under supercritical conditions. Rh-based catalysts have been primarily developed for selective ethanol synthesis or other oxygenates. One downfall for these catalysts is their low catalytic activity which results in the need for high catalyst loadings and more drastic reaction conditions. Coupling this with their high cost and limited availability creates limited commercialization potential of these processes. For all Group VIII metal catalysts, CO conversion to hydrocarbons will be a significant side reaction. It has been observed that the selectivity to oxygenates of Rh-based catalysts is highly dependent on the support, promoter, and metal precursor used.

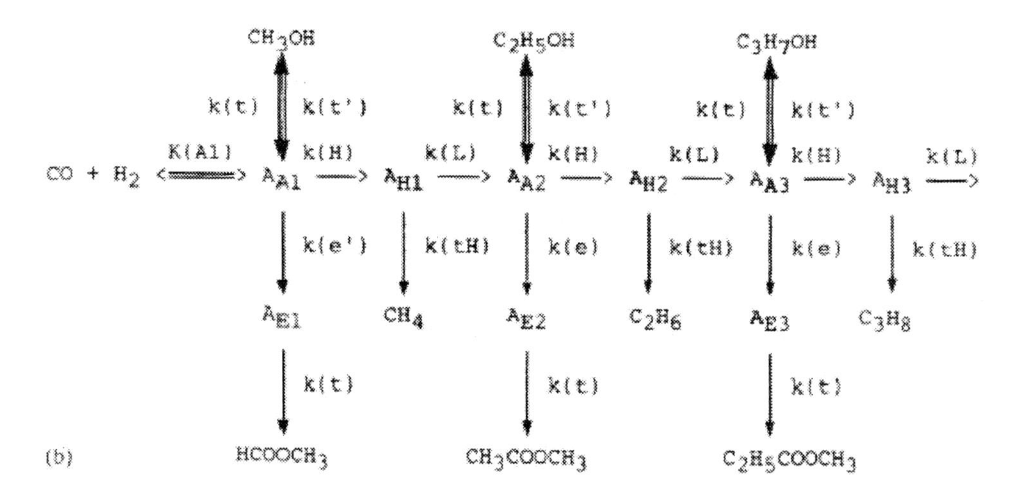

Figure 4. Kinetic Analysis of HAS with Fischer-Tropsch Mechanism

Appendix I Table 5.

Doc #	Title	Date	Author(s)	Company/University Affiliation	Journal / Conference	Synopsis
9542	Synthesis of Higher Alcohols from Syngas over Cu-Co2O3/ZnO, Al2O3 Catalyst	2005	Mahdavi, V.; M.H. Peyrovi, et.al.	Shahid Beheshti University, Tehran Iran; National Iranian Oil Company, Petroleum Research Institute	Applied Catalysis, Elsevier	Fixed bed flow reactor; changes in activity and selectivity resulting from feed and operating conditions; alcohols selectivity greater than 80%
	Mixed Alcohols from Syngas: State of Technology Final Report	2005	Nexant	Nexant	Subcontract to NREL/DOE	State of technology of mixed alcohols and recommendations for areas of focus for research
	Gridley Ethanol Demonstration Project Utilizing Biomass Gasification Technology: Pilot Plant Gasifier and Syngas Conversion Testing	2005	TSS Consultants	City of Gridley, CA	NREL Subcontract Report SR- 51 0-37581	Pearson Technologies Pilot Gasifier; Rice Straw into Ethanol thermochemically; use PSA to remove CO2 from system; Not much data, this is a publicly available report, supposedly there is a more detailed proprietary report
N/A	Production of Mixed Alcohol Fuels	2004	Lucero, Andrew J.	Western Research Institute	DOE workshop on BLG	Update on efforts to use PEFI's modified MoS2 catalyst; compression costs domin-ate op costs; temp selection is critical; target alcohol composition given; Single-pass CO conversions up to 20% to alcohols
11631	Solvent Effects on Higher Alcohols Synthesis Under Supercritical Conditions: a Thermodynamic Consideration	2004	Qin, Zhangfeng; et.al.	Institute of Coal Chemistry, Chinese Academy of Sciences	Fuel Processing Technology, Elsevier	Supercritical HAS, used redlichkwong soave; used various solvents; low ethanol; nice chart of equilibrium constants, standard ethalpies, entropies, and gibbs free energies
9543	Higher-Alcohols Biorefinery: Improvement of Catalyst for Ethanol Conversion	2004	Olson, Edwin S.; et.al.	EERC, North Dakota	Applied Biochem & Biotech	React methanol (from gasification) with ethanol (from biochem) to yield primarily C4 (C3+) alcohols (Guerbet rxn)
	Preliminary Screening - Technical and Economic Assessment of Synthesis Gas to Fuels and Chemicals with Emphasis on the Potential for Biomass- derived Syngas	2003	Spath, P.L.; Dayton, D.C.	National Renewable energy Laboratory (NREL)	Report TP-510-34929	overview of potential for syngas convers-ion to fuels and chemicals; includes hydro-gen, ammonia, methanol, DME, mixed al-cohols, oxosynthesis, MTG, isosynthesis, ethanol fermentation, etc. Very good report
11635	Palladium-based Catalysts for the Synthesis of Alcohols	2003	M. Josefina Perez-Zurita; Cifarelli, M.; et.al.	Universidad Central de Venezuela	Journal of Molecular Catalysis, Elsevier	Catalyst analysis from XRD; high methane and methanol selectivities

(Continued)

Doc #	Title	Date	Author(s)	Company/University Affiliation	Journal / Conference	Synopsis
9527	Synthesis of Higher Alcohols in a Slurry Reactor with CS-promoted Zinc Chromite Catalyst in Decahydrona-phthalene	2003	Xiaolei Sun, G.W. Roberts	North Carolina State	Applied Catalysis, Elsevier	Continuous Slurry Reactor; Cs shifts products away from MeOH, towards Higher Alcohols
9535	Fuel from the Synthesis Gas - the Role of Process Enginee-ring	2003	Stelmachowski, Marek, et.al:	Technical University of Lodz, Poland	Applied Energy, Elsevier	MeOH, FT, and HAS, synthesis, slurry phase reactor; simulation
N/A	Alcohol Synthesis over Pre-Reduced Activated Carbon-supported Molybdenum-Based Catalysts	2003	Dadyburjor, D.B., Li, Xianguo; et.al.	West Virginia University; Ocean University of China	Molecules	formation of alcohols at expense of CO conversion; prereduced rather than sulfided; getting lots of HCs
N/A	Co-Production of Fuel Alcohols & Electricity via Refinery Coke Gasification	2003	Ravikumar, Ravi; and Shepard, Paul	Fluor	Gasification Technologies Conf	Use of Power Energy Fuels Inc's Ecalene process for mixed alcohols synthesis in refinery application (Pet Coke); modeling and economic analysis
9519	The Influence of Clay on K_2CO_3/Co- MoS_2 Catalyst in the Production of Higher Alcohol Fuel	2002	Iranmahboob, Jamshid; Hill, Donald	Uconn, Mississippi State	Fuel Processing Technology, Elsevier	Clay had significant impact on HAS when K2CO3 present; oxygenates selectivity of 70%; testing of feed conditions
9536	Alcohol Synthesis from Syngas over K_2CO_3/CoS/MoS_2 on Activated Carbon	2002	Iranmahboob, Jamshid; Hill, Donald	Uconn, Mississippi State	Catalysis Letters	Experimentation; Effects of Temp, H2:CO, GHSV tested; fixed bed; active chemicals on carbon decreased surface area dram-atically
9528	Advances in Catalytic Synthesis and Utilization of Higher Alcohols	2000	Herman, R.G.	Lehigh Univ.	Catalysis Today, Elsevier	focus on oxygenates, particularly isobu-tanol; slurry phase bed and Double Bed designs; some literature results and op conditions reported; injection of EtOH may promote HAS
	The Economical Production of Alcohol Fuels from Coal-derived Synthesis Gas: Final Report, October 1991 - March 1998	1999		West Virginia University	7 years of research on mixed alcohols from MoS2 catalystsl; Submitted to National Energy extensive work; higher alcohol Technology Lab (NETL),selectivity > 40%; productivity at	

(Continued)

Doc #	Title	Date	Author(s)	Company/University Affiliation	Journal / Conference	Synopsis
					Contract #DE-AC22-91PC91034- 200 g/kg/h; hydrocarbon selectivity 25 < 20%; Use Coal feedstock	
9545	Synthesis of Higher Alcohols - Enhancement by the Addition of Methanol or Ethanol to the Syngas	1999	Lachowska, M.	Polish Academy of Sciences	React.Kinet.Catal.Lett Elsevier	Increase in C3+alcohols using modified methanol catalyst
N/A	Formation of Ethanol and Higher Alcohols from Syngas	1999	Iranmahboob, Jamshid; Donald Hill	M ississippi State University	Mississippi State University, Thesis Report	Talk of biomass; MoS2; Ph. D. Thesis; un-supported (K2CO3/CoMoS2), and suppor-ted on Clay. Mention of methanol removal and adsorption for kinetic driving force
9515	Selective Synthesis of Mixed Alcohols Catalyzed by Dissolved Base-activated Highly Dispersed Slurried Iron	1999	Mahajan, D. et.al.	Brookhaven National Laboratory	Fuel (Elsevier)	Ultrafine catalyst system, lower temps (250C) and pres (<6MPa); up to 95% selectivity (c1-c4 alcohols); and 80% CO Conversion; nanometer iron oxides "NANOCAT "; synthesis remains a problem
9538	Synthesis of Short Chain Alcohols over a Cs-promoted $Cu/ZnO/Cr2O_3$ catalyst	1998	Majocchi, L.; Forzatti, Pio; et.al.	Politecnico di Milano; Snamproget ti SpA	Applied Catalysis, Elsevier	Effect of Temp, CO:H2 ratio, GHSV. Temp is key to Alcohol Productivity, 290 - 315 C are best for C2-C3; HC's minimized
9511	Higher Alcohols from Synth-esis Gas Using Carbon-Supported Doped Molybd-enum -Based Catalysts	1998	Xianguo Li; Dadyburjor, Dady B. et.al.	West Virginia University	Ind. Eng. Chem. Res. (ACS)	Studied catalyst prep parameters and reaction conditions; catalyst promoted w/ K and/or Co
9537	A Kinetic Model for the Synthesis of High-Molecular-Weight Alcohols over a Sulfided Co-K-Mo/C Catalyst	1998	Gunturu, A.K.; Dadyburjor, D.B., et.al	West Virginia University	Ind. Eng. Chem. Res. (ACS)	Kinetics, recycle reactor, temps and feed variables explored; includes rate expressions
9529	Development of a Process for Higher Alcohol Production via Synthesis Gas	1998	Beretta, A.; Tronconi, E. et.al.	Politecnico di Milano; Snamprogetti SpAInd. Eng. Chem. Res.(ACS)	High-temp catalyst; single-stage adiabatic pilot scale reactor, Kinetics develop-ed; targeting oxygenates, ethanol low; simulation of multi-stage adiabatic reactor	

(Continued)

Doc #	Title	Date	Author(s)	Company/University Affiliation	Journal / Conference	Synopsis
9530	High Temperature Calcined K- MoO_3/gamma-Al_2O_3 Catalysts for Mixed Alcohols Synthesis from Syngas: Effects of Mo Loadings	1998	Bian, Guo-zhu, et.al. (Univ. of S&T, China)	University of Sci & Tech, China; Univ of Tokyo	Applied Catalysis, Elsevier	total yields of mixed alcohols decreased but selectivity increased from 3-50%; mainly linear C1-C4, small amt iso-C4
9514	Kinetic Analysis of Mixed Alcohol Synthesis from Syngas over K/MoS_2 Catalyst	1997	Park, T .Y.; Nam, In-sik, et.al.	Pohang Univ. of Sci. & Tech (Postech), Korea	Ind. Eng. Chem Res. (ACS)	K2C O3-promoted-Mo S2; mechanistic kinetics models; mixed alcohol formation maximized around 320C
9518	Synthesis of Mixed Alcohols from CO_2 contained syngas on supported MoS_2 catalysts	1997	Gang, Lu, et.al.	East China University of Sci & Tech	Applied Catalysis, Elsevier	CO2 in feed inhibits higher alcohols; product distribution strongly influenced by CO2 and H2O in feed; high H2:CO ratio used; 0-100 ppm H2S
9524	Development of a Mechanistic Kinetic Model of the Higher Alcohol Synthesis over a Cs-doped Zn/Cr/O Catalyst - 1) Model Derivation and Data Fitting	1996	Beretta, A.; Forzatti, Pio, et.al.	Politecnico di Milano; Snampro-getti SpA	Ind. Eng. Chem. Res. (ACS)	Kinetics model; goal is branched alcohols, not linear because of better octane properties
9525	A Kinetic Model for the Methanol-Higher Alcohol Synthesis from CO/CO2/H2 over Cu/ZnO-based Catalysts Including Simultaneous Formation of Methyl Esters and Hydrocarbons	1994	Breman, B.B.; Beenackers, A.C.C.M.; et.al.	University of Groningen, The Netherlands	Chemical Engineering Science, Pergamon	Overview of past work on modified methanol catalysts and Kinetics; methanol primary product; claims to be most accurate kinetics models to date; substantial HC's produced as unwanted byprod.
9526	Thermodynamics of Higher Alcohol Synthesis	1993	Mawson, S.; Roberts, G.W.; isobutyl; conversions > 90%; et.al. s	North Carolina State	Energy & Fuels	Aspen Plus simulation using RGIBBS; C4 alcohols are dominant product, especially as isobutyl; conversions > 90%; Thermodynamics
11636	SiO2-supported Biometallic Ruthenium Cluster-derived Catalysts for Selectivity and Activity Control in CO Hydrogenation Toward C1-C5 Alcohols	1992	Xiao, F.; Ichikawa, M.; et.al.	Hokkaido University	Journal of Molecular Catalysis, Elsevier	Ruthenium Catalyst;

(Continued)

Doc #	Title	Date	Author(s)	Company/University Affiliation	Journal / Conference	Synopsis
9517	The Thermodynamics of Higher Alcohol Synthesis	1992	Roberts, G.W ., et.al.	North Carolina State	Symposium on NG upgrading II, Division of Petroleum Chem, Inc., ACS	Basically same info as presented in 1993,
9523	An Overview of the Higher Alcohol Synthesis	1992	Smith, K.; Klier, K.	Univ. British Columbia; Lehigh	Symposium on NG upgrading II, Division of Petroleum Chem, inc., ACS	Good brief review on catalyst choices and Operating Conditions at that time, including Rh/SiO2; target for octane seemed to be 70:30 c2+/c1 (required for gasoline blending?)
9521	Non-Stoichiometric Spinel-type Catalysts for the Synthesis of Methanol and Higher Alcohols	1992	Trifiro, F., et.al.	Dipartim ento di Chimica Industriale e dei materiali, Bologna, Italy	Symposium on Octane and Cetane Enhancement Processes for Reducedemissions Motor Fuels, Division of Petroleum Chemistry, ACS	Non-stoichiometric spinel-type solid oxides; again shows propensity of K to shift towards higher alcohols
11632	The Conversion of Synthesis Gas to Higher Oxygenated Fuel on Rh-based Catalysts: Effect of Chemical Additives	1991	Balakos, M.W., Chaung, S. S. C,	University of Akron	Fuel Science and Technology International	Rhodium catalyst, high selectivities to CH4 & C2+ Oxygenates
9522	Higher Alcohol Synthesis	1991	Forzatti, Pio, et.al.	Politecnico di Milano; Snamprogetti SpA	Catal. Rev. - Sci. Eng	Extensive writeup (book chapter?) on status of HAS catalysts; includes kinetics and thermodynamics; and commercial designs from Lurgi, IFP, Snamprogetti
	Production of Oxygenates from Synthesis Gas: A Technology Review for US Dept of Energy for Design of Generic Coal conversion Facilities	1991	Pittsburgh Energy Technology Center	M.W. Kellogg Company	Contract #DE-AC22-91PC89854	Oxygenates overview, including mixed alcohols; discuss dow technology as 85% selectivity and CO perpass conversion of 20-25%; effect of catalyst additives

(Continued)

Doc #	Title	Date	Author(s)	Company/University Affiliation	Journal / Conference	Synopsis
	Classical and Non-classical Routes for Alcohol Synthesis	1991	Herman, R.G.	Lehigh University	Book Chapter, New Trends in CO Activation, L. Guczi, Ed. Elsevier	Impressive summary of state of mixed alcohols technology. Includes detailed kinetics work and experimental results to date
9539	Catalytic Conversion of Synthesis Gas to Methanol and Other Oxygenated Products	1991	Stiles, Alvin, et.al.	University of Delaware	Ind. Eng. Chem. Res.; ACS	Experiments, bench-scale, high aldehyde production based on conditions; benchmark established on pure methanol; alkaline earth & cobalt additives
9531	Kinetic Modelling of Higher Alcohol Synthesis over Alkali-promoted Cu/ZnO and MoS2 Catalysts	1990	Smith, K. Herman, R.; Klier, K.	Lehigh University	Chemical Engineering Science	Kinetics of both modified FT & modified MeOH; different reaction pathways for each
9541	Production of Mixed Alcohols from Synthesis Gas	1990	Uchiyama, S., et.al.	Idemitsu Kosan Co.	Sekiyu Gakkaishi	Comparison of modified FT & modified MeOH Catalysts; includes summary of operating conditions for industry: Halder-Tropsche/Snamprogetti/Enichem, Dow-Union Carbide, IFP/IdemitsuKosan; Lurgi; some economics
9568	Production of Methanol-higher Alcohol Mixtures from Natural Gas via Syngas Chemistry	1990	Coutry, Chaumette, Raimbault, Travers	French Petrol Institute (IFP)	European Applied Research Conf. On Natural Gas	Enichem; Snam Progetti; Haldor Topsoe
	Production of Methanol-higher Alcohol Mixtures from Natural Gas via Syngas Chemistry	1990				Lurgi
	Production of Methanol-higher Alcohol Mixtures from Natural Gas via Syngas Chemistry	1990				IFP & Idemitsu-Kosan
	Production of Methanolhigher Alcohol Mixtures from Natural Gas via Syngas Chemistry	1990				Dow, Union Carbide
	Production of Methanol-higher Alcohol Mixtures from Natural Gas via Syngas Chemistry	1990				C1 Chem Group (Japan), Texaco, others

(Continued)

Doc #	Title	Date	Author(s)	Company/University Affiliation	Journal / Conference	Synopsis
9532	Synthesis of Alcohols from Carbon Oxides and Hydrogen. 4. - Lumped Kinetics for the Higher Alcohol Synthesis over a Zn-Cr-K Oxide Catalyst	1987	Tronconi, E.; Forzatti, P; et.al.	Politecnico di Milano; Snamprogetti SpA	Ind. Eng. Chem. Res. (ACS)	Kinetics; discussions of CO_2 recycle
9512	Mixed Alcohols from Synthesis Gas	1986	Quarderer, George	Dow Chemical	AIChE Spring National Meeting	Very low water, 50:50 MeOH/HAS split; economics provided; reactor designs considered as well; lots of good data
SRI PEP #85-1-4	Mixed Alcohols from Syngas	1986	Nirula, Satish	SRI	PEP Review #85-1-4	Dow, Union Carbide; includes economics; also includes state of art discussion

Basis for Catalyst Selection

Because the focus of this report is thermochemical production of ethanol, a moly-sulfide-based (MoS_2) system promoted with cobalt and alkali metal salts was chosen as the catalyst system because of its ability to produce linear alcohols (as opposed to branched) and its potential for higher ethanol selectivities. This is a form of original Dow/UCC technology.

End Note

[b]Meaning, 1 year or less.

Appendix J. Alcohol Synthesis Kinetics

A more rigorous method for modeling the production of alcohols or other products via synthesis is to use kinetics-based reactions to represent the multiple reactions that occur. Used with the appropriate reactor type, a good kinetics-based model can more accurately predict the effect of varying inlet conditions to the synthesis reactor, especially when unconverted syngas or unwanted co-products (i.e., methanol, hydrocarbons) are recycled to the reactor inlet.

Two major barriers to using kinetics-based models are 1) the need for high-quality experimental data to determine the many parameters required for accurate predictions – assuming that a suitable model can be found for the system of equations – and 2) even with data suitable for estimating model parameters, the resulting model is only truly representative of the catalyst used in the experiments over the range of operating conditions explored. Using the model outside the range of operating conditions introduces increased uncertainty the more removed the estimate is from the experimental conditions used to develop the model. Catalyst performance is very sensitive to many factors that can arise during their production. Two seemingly identical catalysts based on chemical formulation can vary greatly in performance because of preparation techniques and catalyst support characteristics (surface area, crystal structure).

Heterogeneous catalysts such as those used in most synthesis work today, are very sensitive to a number of factors including supports, amounts and types of added promoters or inhibitors that help tune the catalyst to promote desired reactions while inhibiting undesirable reactions. An "optimum" catalyst is usually a compromise of competing and interacting promoters, inhibitors, and supports to give a catalyst with the "best" cumulative properties. When precious metals are used, economics also become a major factor in determining the catalyst formulation.

Very little published data exist for the MoS2-based catalysts that are suitable for developing a reasonable model of the alcohol synthesis together with the very important competing reactions that reduce the desired product yields. Gunturu (Gunturu, et al., 1998; Gunturu, 1997; Gunturu, et al., 1999) published relevant data from thesis work done at the University of West Virginia. The data were used to develop a system of Langmuir-Hinshelwood type reactions to describe a five-reaction lumped kinetic scheme. Recently, researchers in Italy, using the same data, generated a model using the same reaction scheme but with slightly different equations and parameters. The procedure for estimating the parameters is not discussed here, but can be found in detail in Larson, et al. (Larson, et al., 2006) and are presented here in Table 1.

Table 1. Kinetic parameter estimates from Larson, et al. Blank cells do not have values for the corresponding rate equation parameter and rate equation.

Parameter	Methanol	Ethanol	Propanol	Hydrocarbons
A_m, A_e, A_p, Ah	14.6233	3.0518	0.2148	9.3856
E_m, E_e, E_p, Eh	143.472	24.986	89.3328	95.416
n_m, n_e, np, nh	3	1	1	1
K1	7.6393E-9			
K2	0.6785			
K3	0.9987			
K_e		0.7367		
Kp			0.6086	
Kh				1.2472
K_z	0.8359			

The rate equations were created using the typical Langmuir-Hinshelwood kinetic approach found in most catalyst kinetics textbooks. Equation 1 is the gross rate of methanol production. It is the only rate equation in the set that has a reverse reaction component and therefore an equilibrium value. Chemical equilibrium is reached when the partial pressure of methanol is sufficiently high to cause the reverse reaction rate, the dissociation of methanol to CO and H_2, plus the consumption rate of methanol to other products, to equal the forward reaction rate of CO and H2 to make methanol. Given sufficient time in an active catalyst bed, the methanol concentration will reach an equilibrium state. The equilibrium value will include the effects of methanol being consumed to make ethanol or methane as described by equations 4 and 6, respectively.

$$r_{CH3OH}^{gross} = \frac{A_m e^{-(E_m/R)(1/T-1/T_{ep})}\left(\left[\frac{p_{CO}}{p_{CO,ep}}\right]\left[\frac{p_{H2}}{p_{H2,ep}}\right]^2 - \frac{1}{K_{eq}}\left[\frac{p_{CH3OH}}{p_{CH3OH,ep}}\right]\right)}{\left(1+K_1\left[\frac{p_{CO}}{p_{CO,ep}}\right]+K_2\left[\frac{p_{H2}}{p_{H2,ep}}\right]+K_3\left[\frac{p_{CH3OH}}{p_{CH3OH,ep}}\right]\right)^{nm}} \qquad (1)$$

where

$$K_{ep} = \frac{p_{CH3OH,ep}}{p_{CO,ep}\left(p_{H2,ep}\right)^2} \qquad (2)$$

and

$$K_{eq} = \frac{K_a}{K_2 K_{ep}} \tag{3}$$

$$r^{gross}_{C_2H_5OH} = \frac{A_e e^{-(E_e/R)(1/T-1/T_{ep})}\left[\frac{p_{CH_3OH}}{p_{CH_3OH,ep}}\right]}{\left(1+K_e\left[\frac{p_{CH_3OH}}{p_{CH_3OH,ep}}\right]\right)^{ne}} \tag{4}$$

$$r^{gross}_{C_3H_7OH} = \frac{A_p e^{-(E_p/R)(1/T-1/T_{ep})}\left[\frac{p_{C_2H_5OH}}{p_{C_2H_5OH,ep}}\right]}{\left(1+K_p\left[\frac{p_{C_2H_5OH}}{p_{C_2H_5OH,ep}}\right]\right)^{np}} \tag{5}$$

$$r^{gross}_{HC} = \frac{A_h e^{-(E_h/R)(1/T-1/T_{ep})}\left[\frac{p_{CH_3OH}}{p_{CH_3OH,ep}}\right]}{\left(1+K_h\left[\frac{p_{CH_3OH}}{p_{CH_3OH,ep}}\right]\right)^{nh}} \tag{6}$$

The other rate equations describe presumed irreversible reactions of methanol to ethanol, ethanol to propanol, and methanol to methane. Ethanol can react further with 112 and CO to make propanol. Implicit in this set of equations is the simplifying assumption that any products generated other than methanol, ethanol or propanol produce methane even though in higher molecular weight hydrocarbons are observed experimentally. Small amounts of butanol and pentanol, both expected and experimentally observed, have been ignored in the analysis since CO conversions were low resulting in minimal production of increasingly higher alcohols.

Using the rate equations above, plus a rate equation for the water-gas-shift reaction , a system of differential equations and initial conditions can be easily written for a plug flow reactor, as in this report, or for a Continuous Stirred Tank Reactor (CSTR). The solution of the system of ordinary differential equations was programmed into the commercially available equation solver, PolyMath, to evaluate the kinetics over a range of conditions.

Examples of the results given by the model are shown in Figure 1 and Figure 2. In the first case the methanol rate reaches a constant value quickly meaning it is consumed at the same rate it is produced. In the second case methanol was "added" to the reactor inlet to make 1 mol% concentration. CO and 112 inlet flow was reduced equally to keep molar inlet flow the constant. The methanol has a negative rate of production initially and then reaches a constant flow rate at a value slightly below 2000 lb/hr. The ethanol flow rate is significantly higher than the case with no methanol added to the feed.

For the catalyst used to develop this model, the CO conversion is relatively low at 570K and 68 atm pressure. With no methanol recycle to the reactor, the CO conversion is 10.7% with an ethanol yield of 17000 lb/hr. Adding methanol to the reactor inlet increases the CO conversion to 13.5% with an ethanol yield of 25,900 lb/hr. Adding methanol to the synthesis reactor appears to improve the catalyst performance for making ethanol. However, there is insufficient methanol exiting the reactor to meet the inlet demands. At this temperature and pressure, a lower inlet methanol concentration is needed to be sustainable in a process without modifying the process to produce more methanol.

Values of CO conversion and product rates at the reactor outlet for a plug flow reactor at 68, with and without methanol added to the inlet, are given in the kinetic model. The first column is for no methanol added to the inlet stream. The second column has 1 mol% methanol in the inlet. The GHSV is approximately 1000 L-kgcat^{-1}-hr^{-1} for all cases shown, which is significantly less than the experimental condition at which the kinetic parameters were estimated for the kinetic models.

At 570K, adding methanol to the inlet gas causes the rates of ethanol and propanol production to increase significantly with little change in the CO conversion. The methanol equilibrium rate is approximately 2000 Kmol/hr at this temperature. Increasing the temperature to 610K greatly increases the conversion of CO and of the product yields. However, the methane production has increased relative to higher alcohols. The effect of adding methanol at 610K is less pronounced than at 570K. The CO conversion decreases slightly and the alcohol yields are only slightly increased. The methanol equilibrium rate has increased at 610K to approximately 4200 Kmol/hr. In the "methanol added" cases shown, there is insufficient methanol exiting the reactor to supply the inlet after separating the methanol from the other alcohols. A sustainable level would be reached using a lower methanol concentration at the inlet. Higher conversion can also be achieved by decreasing the GHSV although this will require a larger reactor and more catalyst. The optimum combination involves economics to determine when it is infeasible to increase the capital and catalyst costs to increase production.

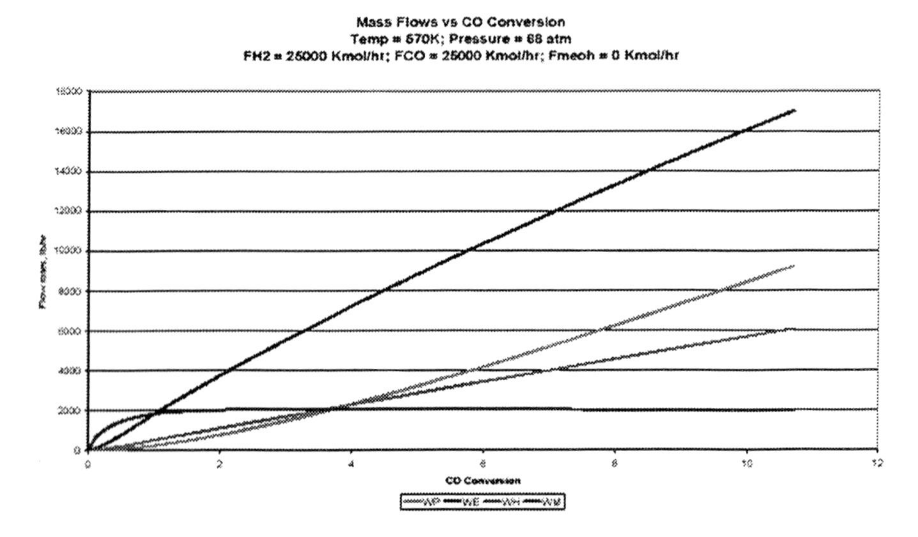

Figure 1. Mass Flow Rates of the Synthesis Reactions vs. CO Conversion in a Isothermal, Plug-Flow Reactor at a WHSV of 1000 L-kgcat^{-1}-hr^{-1}

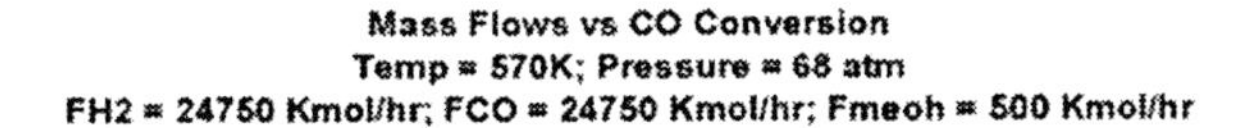

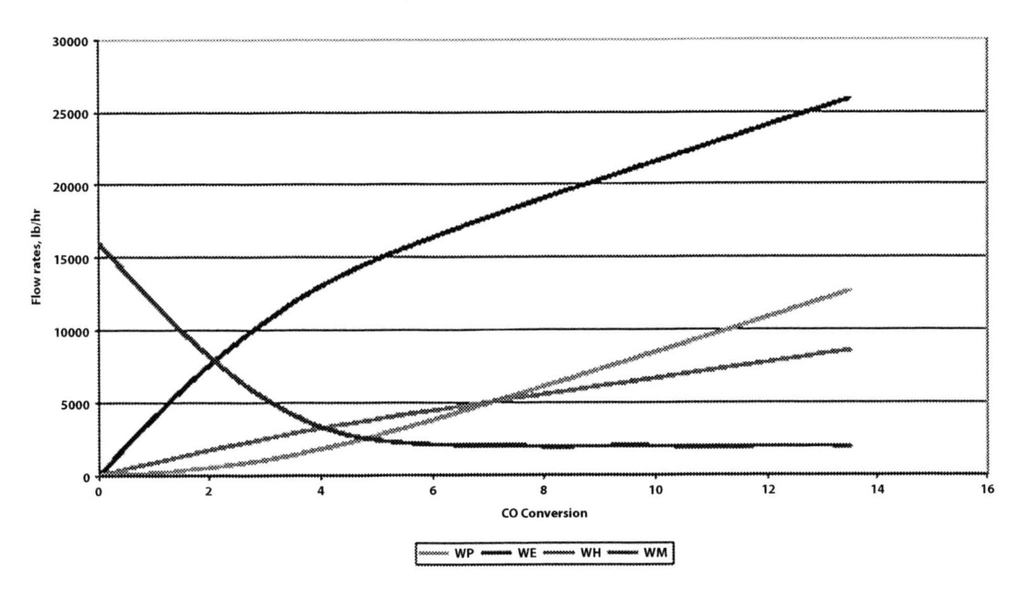

Figure 2. Mass Flow Rates of the Synthesis Reactions vs. CO Conversion in a Isothermal, Plug-Flow Reactor at a WHSV of 1000 L-kgcat^{-1}-hr^{-1}

Table 2. Results From the Kinetic Model at 68 atm Pressure

	MeOH @ inlet = 0 Kmol/hr	**MeOH @ inlet = 500 Kmol/hr**
570K	XCO = 5.67 MeOH = 2023 CH_4 = 3237 EtOH = 9861 PrOH = 21400	XCO = 6.38 MeOH = 1974 CH_4 = 5770 EtOH = 19600 PrOH = 6543
610K	PrOH = 3851 XCO = 27.08 MeOH = 4282 CH_4 = 26400 EtOH = 25600	XCO = 25.94 MeOH = 4197 CH_4 = 28800 EtOH = 28300 PrOH = 23800

The effect of adding methanol on the ethanol yield suggests that other configurations may be more effectively used to maximize ethanol production. Since the methanol production at the reactor exit is limited by chemical equilibrium, the amount of methanol that can be recycled to the reactor is also limited. Instead of recycling only methanol produced from a mixed alcohol catalyst, it may be possible to economically split the main syngas flow with one portion going to a methanol synthesis reactor to make methanol and the remaining fraction going to the mixed alcohol reactor where the methanol is added to the syngas before entering the reactor. It could also be possible to do the reaction in series with the total syngas going through a methanol synthesis catalyst first followed the mixed alcohol catalyst. The methanol catalyst section would need to be sized to give only partial conversion to methanol, contrary to the way these reactors are typically operated when methanol is the desired end product. These alternate designs would require other factors to be considered, such as lower sulfur and CO_2 concentrations for methanol synthesis relative to the mixed alcohol catalyst used.

As more data become available, the kinetic model can be updated to include different catalysts and expanded operation ranges, especially in regards to the amount of methanol and CO_2 that can be fed to the reactor without adversely affecting product selectivity and CO conversion.

REFERENCES

Gunturu, A; et.al. "A Kinetic Model for the Synthesis of High-Molecular-Weight Alcohols over a Sulfided Co-K-Mo/C Catalyst." *Ind. Eng. Chem. Res.* Vol. 37, 1998. pp. 2107 – 2115.

Gunturu, AK; Kugler, EL; Cropley, JB & Dadyburjor, DB; "A Kinetic Model for the Synthesis of High-Molecular-Weight Alcohols over a Sulfided Co-K-Mo/C Catalyst", Chapter 5 in "*The Economical Production of Alcohol Fuels from Coal-Derived Synthesis Gas*", Final Report, Contract # DE-AC22-91PC91034-25, West Virginia University Research Corporation on behalf of West Virginia University, March 1999.

Gunturu, AK;. "*Higher-Alcohol Synthesis from Carbon Monoxide and Hydrogen: Kinetic Studies Over MoS2-Based Catalysts*", M.S. Thesis in Chemical Engineering, West Virginia University, Morgantown, West Virginia, 1997.

Larson, ED; Consonni, S; Katofsky, RE; Iisa, K; Frederick, J;. "A Cost-Benefit Assessment of Gasification-Based Biorefining in the Kraft Pulp and Paper Industry, Volume 2, Detailed Biorefinery Design and Performance Simulation", Final Report under contract DE-FC26- 04NT42260 with the U.S. Department of Energy and with cost-sharing by the American Forest and Paper Association, December 2006.

REPORT DOCUMENTATION PAGE

Form Approved
OMB No. 0704-0188

The public reporting burden for this collection of information is estimated to average 1 hour per response, including the time for reviewing instructions, searching existing data sources, gathering and maintaining the data needed, and completing and reviewing the collection of information. Send comments regarding this burden estimate or any other aspect of this collection of information, including suggestions for reducing the burden, to Department of Defense, Executive Services and Communications Directorate (0704-0188). Respondents should be aware that notwithstanding any other provision of law, no person shall be subject to any penalty for failing to comply with a collection of information if it does not display a currently valid OMB control number.

PLEASE DO NOT RETURN YOUR FORM TO THE ABOVE ORGANIZATION.

1. REPORT DATE *(DD-MM-YYYY)*
April 2007

2. REPORT TYPE
Technical Report

3. DATES COVERED *(From - To)*

4. TITLE AND SUBTITLE
Thermochemical Ethanol via Indirect Gasification and Mixed Alcohol Synthesis of Lignocellulosic Biomass

5a. CONTRACT NUMBER
DE-AC36-99-GO10337

5b. GRANT NUMBER

5c. PROGRAM ELEMENT NUMBER

6. AUTHOR(S)
S. Phillips, A. Aden, J. Jechura, D. Dayton, and T. Eggeman

5d. PROJECT NUMBER
NREL/TP-510-41168

5e. TASK NUMBER
BB07.3710

5f. WORK UNIT NUMBER

7. PERFORMING ORGANIZATION NAME(S) AND ADDRESS(ES)
National Renewable Energy Laboratory
1617 Cole Blvd.
Golden, CO 80401-3393

8. PERFORMING ORGANIZATION REPORT NUMBER
NREL/TP-510-41168

9. SPONSORING/MONITORING AGENCY NAME(S) AND ADDRESS(ES)

10. SPONSOR/MONITOR'S ACRONYM(S)
NREL

11. SPONSORING/MONITORING AGENCY REPORT NUMBER

12. DISTRIBUTION AVAILABILITY STATEMENT
National Technical Information Service
U.S. Department of Commerce
5285 Port Royal Road
Springfield, VA 22161

13. SUPPLEMENTARY NOTES

14. ABSTRACT *(Maximum 200 Words)*
This process design and technoeconomic evaluation addresses the conversion of biomass to ethanol via thermochemical pathways that are expected to be demonstrated at the pilot level by 2012.

15. SUBJECT TERMS
process design; thermochemical; pilot; cellulosic; ethanol

16. SECURITY CLASSIFICATION OF:			17. LIMITATION OF ABSTRACT	18. NUMBER OF PAGES	19a. NAME OF RESPONSIBLE PERSON
a. REPORT	b. ABSTRACT	c. THIS PAGE			
Unclassified	Unclassified	Unclassified	UL		19b. TELEPHONE NUMBER *(Include area code)*

In: Ethanol Biofuel Production
Editor: Bratt P. Haas

ISBN: 978-1-60876-086-2

Chapter 4

WHEY TO ETHANOL: A BIOFUEL ROLE FOR DAIRY COOPERATIVES?

United States Dept. of Agriculture

ABSTRACT

Pertinent information regarding whey-to-fuel ethanol production is explored and reviewed. A potential of producing up to 203 million gallons of fuel ethanol from whey in 2006 was estimated, and dairy cooperatives could have a share of 65 million gallons. Two whey-ethanol plants are currently operated by dairy cooperatives, producing a total of 8 million gallons a year. Successful operations of the plants since the 1980s indicate that (1) fuel ethanol production from whey is technically feasible, (2) whey-to-fuel ethanol production technologies and processes are mature and capable of being adopted for commercial operations, and (3) producing fuel ethanol from whey is economically feasible. However, in this era of whey products' price uncertainties, a key consideration in assessing the feasibility of a new whey-ethanol venture should be the valuation of the opportunity cost of whey as feedstock for fermentation. A new whey- ethanol plant probably should have an annual production capacity of at least 5 million gallons of ethanol. Some historical lessons on the pitfalls to avoid are summarized.

Key Words: Whey, whey permeate, permeate mother liquor, lactose, ethanol, dairy cooperatives.

ACKNOWLEDGMENTS

The author would like to thank Mr. John Desmond of the Carbery Group and Dr. M. Clark Dale of Bio-Process Innovation, Inc., for providing information on the whey- ethanol production processes. For their cooperation in sharing information about whey- ethanol plant operations, the Dairy Farmers of America and Land O'Lakes are also gratefully acknowledged. Credits are also due to many people who were consulted during the course of this study.

Mention of company and brand names does not signify endorsement over other companies' products and services.

HIGHLIGHTS

An estimated 90.5 billion pounds of whey was generated as a byproduct of cheese production in 2006. Besides the liquid carrier, the composition of whey is approximately 0.3 percent butterfat, 0.8 percent whey proteins, 4.9 percent lactose, and 0.5 percent minerals. Cumulatively, there were 4.4 billion pounds of lactose contained in the whey produced that year.

Whey may be made into many products with various processes and technologies. Condensed whey, dry whey, dry modified whey, whey protein concentrate and isolates, as well as lactose (crystallized and dried) are the often cited whey products. There are many other secondary and tertiary products that can be derived from whey, but the volume of whey used in these products is relatively small.

Whey products produced in 2006 contained an estimated total of 1.9 billion pounds of lactose. Therefore, about 2.5 billion pounds of surplus lactose were unaccounted for by whey products. This vast amount of surplus lactose could be fermented to produce an estimated 203 million gallons of ethanol, assuming complete consumption of lactose in fermentation and ethanol conversion efficiency at 100 percent of the theoretical yield. Dairy cooperatives' share of the whey-ethanol potential could be 65 million gallons.

There are two industrial-scale whey-ethanol plants in the United States, at Corona, Calif., and Melrose, Minn. Both began operation in the 1980s and are currently owned and operated by dairy cooperatives. Together they produce 8 million gallons of fuel ethanol a year.

The whey-to-ethanol plant commissioned in 1978 by Carbery Milk Products Ltd. of Ireland is believed to be the first modern commercial operation to produce potable (drinkable) alcohol. Starting in 1985, it has produced fuel ethanol as well. The Carbery process developed by the company has been adopted by plants in New Zealand and the United States. New Zealand started using fuel ethanol produced from whey in August 2007.

All ethanol production processes share some basic principles and steps. Whey permeate from protein ultrafiltration is concentrated by reverse osmosis to attain high lactose content. Lactose is fermented with some special strains of yeast. Once the fermentation is completed, the liquid (beer) is separated and moved to the distillation process to extract ethanol. This ethanol is then sent through the rectifier for dehydration and then denatured. The effluent (stillage and spent yeast) may be discharged to a treatment system, digested for methane gas, sold as feed, or further processed into food, feed or other products.

To be economically viable, a dehydration plant (and by inference, an ethanol plant) needed to have a minimum daily capacity of 60,000 liters of ethanol (about 15,850 gallons a day or five million gallons a year), according to a 2005 New Zealand report. The estimated "at-gate" cost (operating and capital service costs) of producing ethanol from whey permeate at maximum technical potential, with a level of uncertainty of +/- 20 percent, was N.Z. $0.6-0.7 per liter. Using a currency exchange rate of N.Z. $1 = U.S. $0.7, the estimated cost translated to U.S. $1.60-1 .85 per gallon. This estimate is similar to the costs quoted by sources in the United States: about $1 per gallon of operating cost and a capital service cost that is predicated on the capital cost ranging from $1.50 to $4 per annual gallon for a commercial operation, depending on the scale of the plant. The estimated operating cost assumes that whey permeate used in ethanol fermentation is a free (no cost) feedstock. Capital cost is the cost of the plant construction project.

There is an opportunity cost of lactose for ethanol fermentation only if there are competing uses of the same lactose, such as manufacturing dry whey, lactose, or other whey products. If there is no such competition, then the whey permeate somehow has to be disposed of and the opportunity cost of lactose for ethanol fermentation is likely to be zero or even negative.

It takes 12.29 pounds of lactose to produce a gallon of ethanol, if the lactose is completely consumed in fermentation and ethanol conversion efficiency is 100 percent of the theoretical yield. For every $0.01 net lactose value (price of lactose net of processor's cost), the feedstock cost for fermentation would be $0.1229 per gallon of ethanol. If lactose consumption is less than complete in fermentation and ethanol conversion efficiency is less than 100 percent of the theoretical yield, then more than 12.29 pounds of lactose is required to produce a gallon of ethanol and the feedstock cost would be higher.

Whether it is economically feasible to produce ethanol from whey permeate is determined by the balance of the production costs and the expected revenues. Net returns from the ethanol enterprise should be measured against the profitability of making other whey products or of other uses, to determine whether ethanol production is a more worthwhile undertaking. A further consideration should be which of the whey enterprises fit best with a cooperative's overall business strategy.

The fact that the two whey-ethanol plants have been in operation for more than 20 years is an indication that (1) fuel ethanol production from whey is technically feasible, (2) whey-to-fuel ethanol production technologies and processes are mature and capable of being adopted for commercial operations, and (3) producing fuel ethanol from whey is economically feasible.

In assessing the feasibility of a new whey-ethanol plant, the cost of whey permeate as feedstock needs to be carefully evaluated in this era of whey products' price uncertainties. Other important factors to consider besides the feedstock cost are (1) an appropriate plant scale that would minimize capital cost and the cost of assembling feedstock, (2) an appropriate technology and processes that would minimize operating cost, (3) best alternatives for using and/or disposing of the effluent, (4) ethanol price, and (5) various government production incentives.

Dairy cooperatives are certainly well-positioned to coordinate whey assembly for ethanol production. However, in view of the current high and unsettled dry whey products prices, there are great uncertainties concerning the long-term development of the whey-ethanol production enterprise.

There was a very high attrition rate of fuel ethanol plants during the decade of 1980s. Experiences of that period provide some lessons that may be relevant to future commercial whey-ethanol development. To be successful, a fuel ethanol plant should have proper technology selection, proper engineering design, adequate research support, credible feasibility studies, adequate financing; and personnel with technical and managerial expertise in the biochemical process.

INTRODUCTION

A total of 90.5 billion pounds of whey was estimated to have been generated as a byproduct of cheese production in 2006, comprising about 85.8 billion pounds of sweet whey and 4.7 billion pounds of acid whey (Table 1, *next page*). A general rule of thumb is that the volume of sweet whey is about nine times the volume of cheese produced and the acid whey volume is about six times that of cottage cheese. Over the last 5 years, from 2001 to 2006, the volume of whey increased by 15 percent, commensurate with the increases in the production of cheeses.

The composition of whey varies with the components in milk that is used for making cheese, the variety of cheese made, and the cheese-making process employed. Whey contains approximately 0.3 percent butterfat, 0.8 percent whey proteins, 4.9 percent lactose, and 0.5 percent minerals (*Wisconsin Center for Dairy Research*).

Butterfat is traditionally of high value, and most plants separate it for use as an ingredient for further processing. The remaining whey may be made into various products by using an array of processes and technologies, or is otherwise disposed of (Table 1 and Figure 1, *page 3*).

Whey can be condensed or concentrated, dried, fermented, delactosed, demineralized, and deproteinated. It is adaptable to ultrafiltration, reverse osmosis, ion exchange, electrodialysis, and nanofiltration (*Kosikowski, et al*).

The main whey products are dry products: dry whey, lactose, and whey protein concentrate (Table 1 and Figure 1). These whey products are storable for later distribution over a wide area, even internationally. Condensed whey also uses a significant amount of whey, but the market is limited due to its wet form.

There are many other secondary and tertiary products that can be derived from whey (*Kosikowski, et al*). However, the volume of whey used in these products is relatively small (*Yang, et al*).

While whey products have found wider uses in recent years and of late have become valuable commodities (table 2 and sidebar, page 4), making these products was originally considered a lower-cost, last- resort alternative to dumping surplus whey.

Most of the components of whey can quickly deplete oxygen levels in natural water systems (*Hamilton*). The biochemical oxygen demand (BOD) of whey is about 3.5 pounds per 100 pounds of whey or 35,000 ppm, and its chemical oxygen demand (COD) is about 68,000 ppm (*Webb, et al*). Such high levels of pollutants make disposing of whey problematic.

Methods of disposing of surplus whey include animal feeding, land spreading, or discharging it after treatment for BOD reduction. There are also some recent cases of feeding whey to anaerobic digesters to produce methane gas (*Dairy Facts*).

Animal feeding and land spreading have limitations (*Cotanch, et al; Wendoff; Kosikowski, et al*). Continuous land disposal of cheese whey can endanger the physical and chemical structure of the soil, decrease the crop yield, and lead to serious water pollution problems (*Belem, et al*).

Treating whey for BOD reduction before discharging it is costly. As an exercise to evaluate the cost, cursory searches on the Internet selected 20 sewage districts in as many States (not a random sample) that posted clearly discernible sewage rates on volume, BOD, etc., for 2006-07. Average volume charge was $2.50 per 1,000 gallons of sewage (3 cents per hundredweight) discharged, and average BOD surcharge was $0.27 per pound of BOD that was above a basic level, usually 200-300 ppm. In addition, some jurisdictions also had surcharges on COD and other pollutants.

Table 1. Fluid Whey, and Whey and Modified-Whey Products Produced, 2001 -2006, United States

	2001	2002	2003	2004	2005	2006
Estimated fluid whey volume[1]:	*Billion pounds*					
Sweet type	74.3	76.9	77.0	79.9	82.3	85.8
Acid type	4.5	4.5	4.6	4.7	4.7	4.7
Total	78.8	81.4	81.6	84.6	87.0	90.5
Whey and modified-whey products:	*1,000 pounds*					
Condensed whey, solids, sweet type, human	81,484	108,250	114,656	91,227	79,247	106,919
Dry whey	1,045,655	1,115,321	1,085,165	1,034,898	1,040,692	1,100,346
Reduced lactose and minerals	129,245	124,670	84,110	84,893	98,371	91,596
Lactose	519,161	563,110	613,976	665,621	713,975	738,656
Whey protein concentrate	336,221	313,239	357,944	355,854	383,926	427,724
Whey protein isolates[2]			22,333	27,677	27,595	30,673
Whey solids in wet blends, animal[3]	39,851	37,656				

Sources: *Dairy Products, Annual Summary*, USDA National Agricultural Statistics Service, selected years, unless otherwise specified.

[1] Estimated at 9 times cheese production for sweet whey and 6 times cottage cheese for acid whey.

[2] New data series started with the year 2003. (*Dairy Products*, October 4, 2005).

[3] Not shown when fewer than three reported or individual plant operations could be disclosed.

These various charges highlight the high cost of surplus whey disposal. Making whey products reduces the surplus whey volume, saves on the cost of disposing of whey, and has the prospect of breaking even or making profit in whey plant operations. Thus, it is important for the industry to find new ways to use more whey.

Advances in membrane and filtration technology since the late 1970s enable processors to "harvest" whey proteins, which are of high nutritional value. In recent years, whey proteins have become popular for use in fortifying more and more foods, beverages, infant formulas, and nutraceuticals. The growth in demand has pushed up whey protein concentrate production by 20 percent in 3 years since 2003 and whey protein isolates by 37 percent (Table 1).

Harvesting whey proteins still poses the problem of dealing with whey permeate, which retains most of the lactose and other solids. Whey permeate may be dried for feed or food uses, but the largest volume is used to produce lactose. However, lactose has somewhat limited application in food products because of its low digestibility and poor solubility: it is prone to crystallization (*Audic, et al; Alexander, et al*). In addition, producing lactose has a leftover product-permeate mother liquor, which contains about 60 percent lactose (dry basis) - that still needs to be disposed of (*Dale, et al*).

The issue of profitably handling the large volume of surplus whey remains. Producing ethanol by fermenting lactose contained in whey, whey permeate, and permeate mother liquor may be a promising alternative. In common usage, ethanol is often referred to simply as alcohol.

Table 2. Average Annual Prices of Whey Products, Carlot or Trucklot Quantities in Bulk Packages, 2001-2006, and Monthly Prices Since 2006

Year	Whey powder, edible nonhygroscopic (Central)	Lactose, edible (Central & West)	Whey protein concentrate, edible 34% protein (Central & West)
		Dollars per pound	
2001	0.2777	0.2090	0.7777
2002	0.1971	0.2042	0.5205
2003	0.1684	0.2094	0.4968
2004	0.2395	0.2262	0.5869
2005	0.2781	0.2012	0.8430
2006	0.3425	0.3333	0.6981
Month			
2006			
January	0.3482	0.2427	0.8004
February	0.3529	0.2492	0.7524
March	0.31 93	0.2500	0.6825
April	0.2875	0.2678	0.6144
May	0.2789	0.2816	0.5990
June	0.2811	0.2873	0.5800
July	0.2901	0.3328	0.5935
August	0.31 71	0.3438	0.6209
September	0.3599	0.3628	0.6703
October	0.4058	0.4139	0.7468
November	0.4308	0.4392	0.8295
December	0.4388	0.5288	0.8869
2007			
January	0.5096	0.5430	1.0012
February	0.6788	0.6062	1.1784
March	0.7768	0.6681	1.3506
April	0.7807	0.9227	1.4801
May	0.7376	0.9370	1.5500
June	0.7385	0.9273	1.6210
July	0.6743	1.0353	1.6460

Source: *Dairy Market News*, USDA Agricultural Marketing Service.

The best known, first commercially operated whey-to-ethanol plant was commissioned in April 1978 by Carbery Milk Products Ltd. of Ireland to produce potable (drinkable) alcohol (*Sandbach*). Since 2005, the company has been suppling ethanol made from whey to an oil firm for E85 and E5 blends (*The Maxol Group; Irish Examiner.com*).

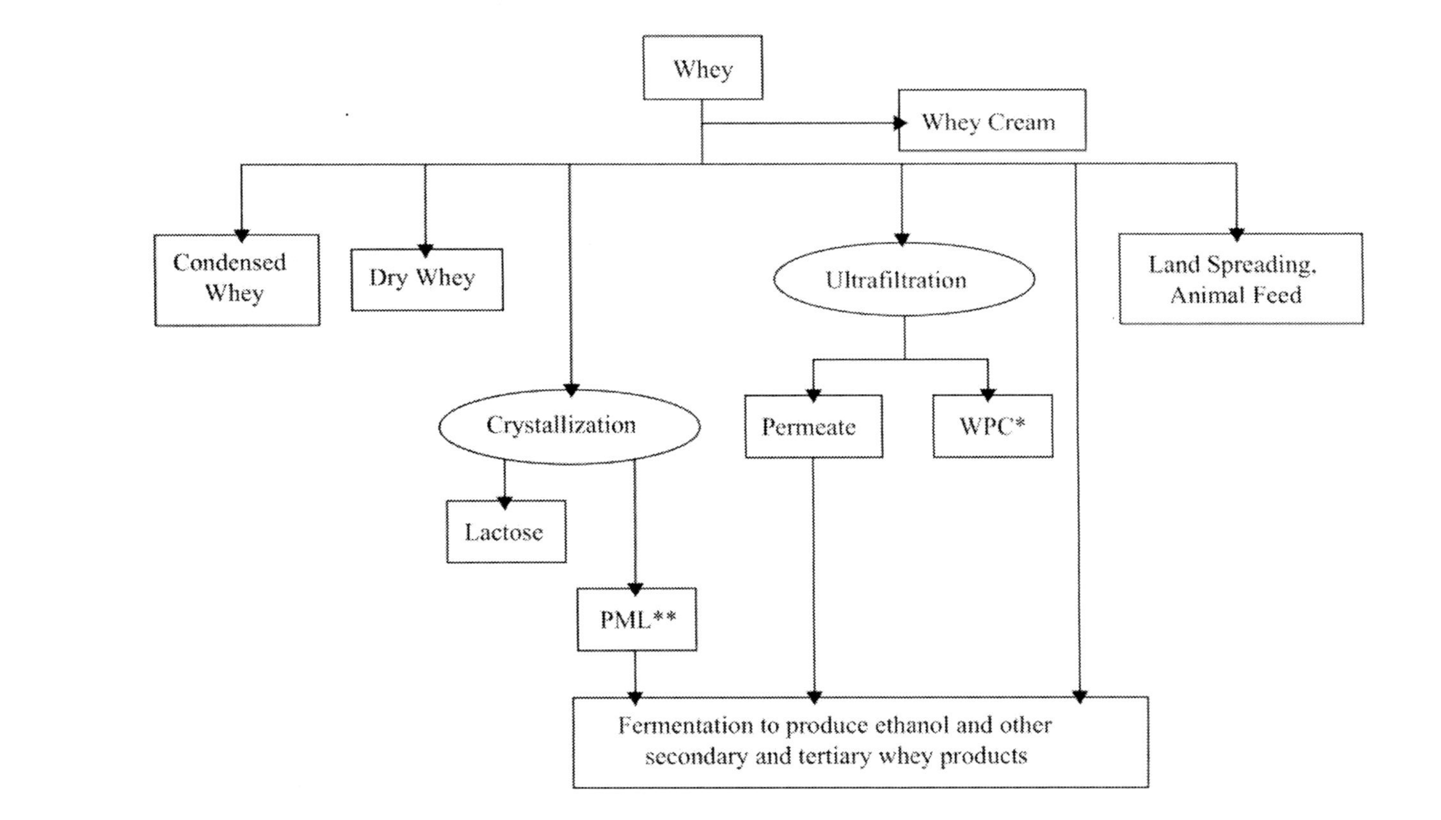

*WPC: Whey protein concentrate

** PML: Permeate mother liquor, which contains 60% lactose, dry basis

Figure 1. Major products and uses derived from whey (Audic, et al. (Figure 2); Dale, et al. (Figure B.1))

HIGH PRICES FOR DRY DAIRY PRODUCTS DUE TO DIVERSE REASONS

Recent high prices of dry dairy products are due to the following factors (USDA *Economic Research Service; OECD- FAO*):

- European Union (EU) agricultural reforms in 2003 reduced the incentives for producing butter and dry nonfat milk, thus shifting more milk solids into cheese.
- Worldwide growth in both cheese consumption and an array of milk protein concentrates (MPC) also reduces the amount of milk protein that might otherwise be made into nonfat dry milk.
- Weather-related problems in Australia reduce the amount of dry dairy products available for export.
- A strong world economy spurs growth in world demand for dairy products.
- Lower exchange rates of U.S. dollar in recent years improve the competitiveness of U.S. dairy products.

Policy changes in the EU and worldwide growth in cheese and MPC demand represent fundamental changes, while weather problems, economic growth, and exchange rates are to some extent cyclical.

The Carbery process has been adopted by plants in New Zealand and the United States.

Internationally, the most notable whey-to-ethanol producer is Anchor Ethanol Ltd, which is wholly owned by the New Zealand dairy cooperative Fonterra. It operates three ethanol plants with an annual total production of about 5 million gallons, and claims to be the largest ethanol producer in the world that uses whey (from casein plants in this case) as feedstock. The ethanol has been used for food, beverage and industrial applications (*Anchor Ethanol; Mackle*). Beginning on August 1, 2007, the company's fuel ethanol has become available (*Farmnews for NZ Farmers*). It has been blended with gasoline and sold commercially as E10.

In the United States, there are two industrial- scale plants that produce fuel ethanol from whey. Both are currently owned and operated by dairy cooperatives. Dairy Farmers of America (DFA) operates the Corona, Calif., plant through its Golden Cheese Company subsidiary. The Melrose, Minn., plant operated by Land O'Lakes is part of a cheese production joint venture between Land O'Lakes and DFA.

The Corona plant has been in operation since 1985, except for a hiatus of more than a year from 1998 to 2000. It is located on the same premises as the cheese plant. The ethanol plant was originally licensed to use the Carbery process for producing ethanol from whey permeate. It has a production capacity of 5 million gallons of ethanol per year. Proprietary yeast is propagated at the plant, and the cells are recycled and reused several times. Fermentation takes place in eight batches, and each batch requires slightly more than 24 hours to complete. In the distillation column the ethanol is concentrated to 190 proof. It is then dehydrated to 200 proof for fuel ethanol and is denatured before shipping.

Dairy Farmers of America's ethanol-from-whey plant in Corona, California. (Photograph courtesy of Dairy Farmers of America)

Table 3. Estimated Volume of Lactose in Whey Products, 2006

Item	Product	Lactose	
	Million lbs	*Percent*[1]	*Million lbs*
Lactose in sweet whey	85,809.0	4.9	4,205
Lactose in cottage cheese (acid) whey[2]	4,651.9	4.9	228
Total lactose volume (estimated)			4,433
Lactose used in whey products (estimated): Condensed whey, solids content[3]	106.9	77.5	83
Dry whey products			
Dry whey, Total	1,100.3	74.4	819
Reduced lactose & minerals[4]	91.6	71.3	65
WPC, 25.0-49.9% protein[5]	297.5	51.0	152
WPC, 50.0-89.9% protein[6]	130.3	5.0	7
Whey protein isolates, 90.0% and higher	30.7	1.0	0
Lactose[7]	738.7	99.0	731
Total lactose used in whey products			1,857
Lactose unaccounted for by whey products			2,576

[1] Adopted from *Wisconsin Center for Dairy Research*, unless otherwise specified.
[2] Cottage cheese whey contains 4.9% lactose (*Kosikowski, et al*, p. 427). Other references tend to report lower lactose content.
[3] Percentages among solids in dry whey, not counting moisture. Condensed whey at 20% solids is estimated to contain 15.5% lactose.
[4] Average composition of reduced-lactose whey and reduced-mineral whey.
[5] Uses composition for WPC-34.
[6] Uses composition for WPC-80.
[7] Uses composition for food-grade lactose (*Chandan*).

(DFA recently announced the closure of the Corona facility. The plant would operate at a reduced capacity beginning August 31 and cease production of American block cheese and whey products by December 31, 2007 (*DFA*).)

The Melrose plant began operation in 1982 and currently is producing about 3 million gallons a year. It is located a couple hundred feet away from the cheese plant. All products and utilities are brought to the ethanol plant via a pipe rack from the cheese plant. The technology was originally developed by Kraft, and yeast propagation is also proprietary. It uses a fed-batch fermentation system that ferments seven batches a day. Whey permeate is sent to fermentation tanks and inoculated with yeast. After fermentation, water and ethanol are separated from the lactose distillers solids using an old whey evaporator. The water and ethanol mixture is then sent to a distillation column and a molecular sieve for concentration and dehydration. (Note: In a fed-batch system, substrate is fed into the fermentation tank at constant intervals, while effluent is removed continuously (*Roehr*).)

Both plants concentrate lactose in the whey permeate to more than double its natural strength before fermentation. No other pretreatment on the feedstock is required. Lactose is almost completely consumed in the fermentation process. The resulting beer contains a level of ethanol that is required for efficient distillation and dehydration. Most effluent from ethanol production is sold for animal feed.

The ethanol production of 8 million gallons from these two plants accounts for a minor portion (less than 1 percent) of total U.S. annual fuel ethanol production. The total surplus lactose volume as calculated below shows that potential exists to produce up to 203 million gallons of ethanol a year from whey. This study attempted to ascertain the feasibility of expanding whey-ethanol production by reviewing the potential volume of ethanol from whey sources, the current processes of whey permeate to ethanol conversion, the economics of producing fuel ethanol from whey permeate, and the organization of whey-ethanol plant operation in which dairy cooperatives may play a role.

Table 4. Comparison of Lactose Volumes that May Be Used for Ethanol Production, 2003-2006

Item	2003	2004	2005	2006
	Million lbs			
Total lactose volume (estimated)	4,000	4,142	4,266	4,433
Lactose used in whey products (estimated):				
Condensed whey, solids content	89	71	61	83
Dry whey products				
Dry whey, Total	807	770	774	819
Reduced lactose & minerals	60	61	70	65
WPC, 25.0-49.9% protein	139	139	141	152
WPC, 50.0-89.9% protein	4	4	5	7
Whey protein isolates, 90.0% and higher	0	0	0	0
Lactose	608	659	707	731
Total lactose used in whey products[1]	1,707	1,703	1,759	1,857
Lactose volume unaccounted for that could be used for ethanol production	2,293	2,439	2,506	2,576
	Million gallons			
Potential volume of ethanol production (estimated)	182	195	199	203
Estimated actual production in 2006				8

[1] Items may not add to total due to rounding.

POTENTIAL VOLUME OF ETHANOL FROM WHEY SOURCES

Ethanol from whey is produced by fermenting the lactose contained in whey, whey permeate, or permeate mother liquor. Therefore, the potential volume of ethanol production from whey feedstock depends on the available volume of surplus lactose that is not used in whey-derived products.

Volume of Surplus Lactose

The 90.5 billion pounds of whey generated by the cheese industry in 2006 contained an estimated 4,433 million pounds of lactose (Table 3, *opposite*). An estimated total of 1,857 million pounds, or 42 percent of available lactose, was used in these main whey products: condensed whey, dry whey, reduced lactose and minerals whey, whey protein concentrates, whey protein isolates, and lactose.

Hence, an estimated 2,576 million pounds of lactose was unaccounted for in 2006 by these whey products. Some of this unaccounted for volume could have been in secondary and tertiary whey products and other whey or lactose-derived products. Therefore, it may be reasonable to estimate that there was about 2.5 billion pounds of surplus lactose in 2006. This is the amount of lactose that may be available for ethanol production.

Potential Ethanol Volume

Theoretically, 1 pound of lactose would yield 0.538 pound of ethanol. Therefore, the potential volume of ethanol production from surplus lactose in 2006 may be estimated at about 203 million gallons.

In the same way, potential ethanol volumes from surplus lactose were estimated for previous years: 182 million gallons in 2003, 195 million gallons in 2004, and 199 million gallons in 2005 (Table 4, *opposite*). The 2006 volume was a 12-percent increase from 2003.

In 2006, the two ethanol plants operated by dairy cooperatives together produced 8 million gallons of ethanol. That still left 195 million gallons as untapped potential.

Implicit in the estimation of potential ethanol volume is the premise that use of lactose in food, feed, industrial, and other applications should take precedent, and ethanol production is the last-resort use of whey and lactose. As will be seen later in the discussion of the economics of whey-ethanol production, every cent of net lactose value (price of lactose net of processor's cost) would increase feedstock cost of ethanol fermentation by at least 12.29 cents per gallon of ethanol. Last-resort use of whey and lactose in ethanol production would keep the feedstock cost as low as possible.

Share of Dairy Cooperatives

Dairy cooperatives produced 40 percent of the Nation's natural cheese in 2002 (*Ling*). Presumably, they also accounted for 40 percent of the whey generated. In that same year, dairy cooperatives produced 1.1 billion pounds of dry whey products, or 52 percent of U.S. total volume, in 28 dry whey plants that they operated.

Using these same ratios for 2006, dairy cooperatives would account for an estimated 800 million pounds of lactose that was not used in whey products. Of this amount, about 100 million pounds was used to produce 8 million gallons of ethanol by the two plants operated by dairy cooperatives. The remaining 700 million pounds of lactose represents a potential volume of 57 million gallons of ethanol.

PROCESSES OF WHEY PERMEATE TO ETHANOL CONVERSION

The first patent for the use of whey in ethanol production (*U.S. Patent no. 2,183,141*) was granted in 1939 (*Murtagh*).

The best known commercial process of producing ethanol from fermenting whey is the Carbery process, which was developed in Ireland for making potable alcohol and was later adopted for industrial alcohol and fuel ethanol as well.

In the United States, there is the proprietary process used by the plant in Melrose, Minn. In addition, research efforts have culminated in the development of some successful processes (*Dale, et al; U.S. Department of Energy (DOE), Office of Industrial Technologies, Energy Efficiency and Renewable Energy; Bio-Process Innovation, Inc.*). Many novel processes also have been seen in media reports and research literature.

The Basics

Ethanol production processes may vary between plants, but they all share some basic principles and steps (Figure 2).

After whey protein has been harvested from whey by ultrafiltration, the remaining permeate is concentrated by reverse osmosis to attain higher lactose content for efficient fermentation.

Lactose in whey permeate is fermented with some special strains of the yeast *Kluyveromyces marxianus* that are efficient in fermenting lactose. The yeast is added to the fermenting substrate and pumped to the fermentation vessels.

Once the fermentation has been completed, yeast is separated from the fermented substrate, and the remaining liquid (beer) is moved to the distillation process to extract ethanol. This ethanol is then sent through the rectifier for dehydration. If the resulting anhydrous ethanol is intended for fuel, it is denatured by adding gasoline to prevent misuse.

The effluent—the remaining liquid after ethanol has been removed from the beer (stillage) and the biomass (spent yeast)—may be discharged to a treatment system, digested for methane gas, sold as feed, or further processed into food, feed, or other products. (For a concise description of the manufacture of ethanol from whey, see *Hamilton*.)

Beyond the basics, there are many variations of whey-ethanol production processes. Two with available current information are the Carbery process and the processes offered by the Bio-Process Innovation, Inc. They are described in Appendix I and Appendix II, respectively.

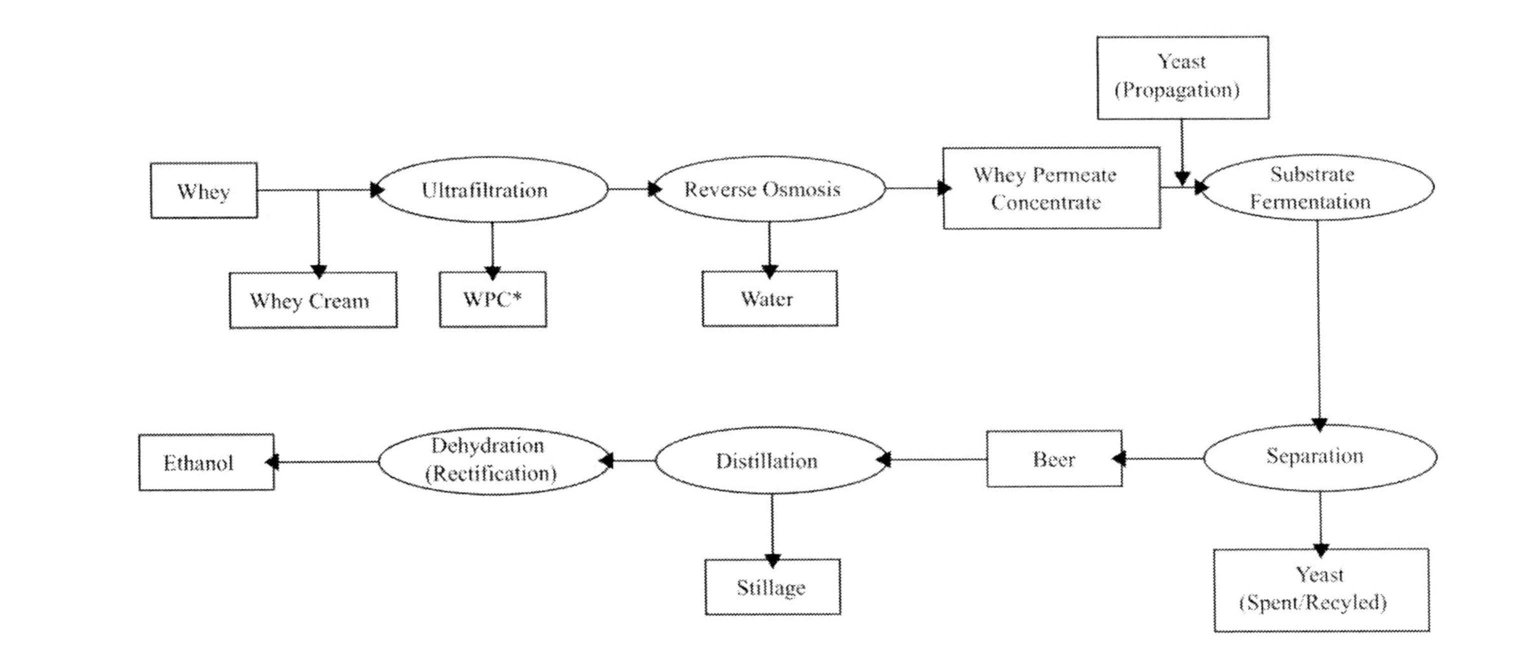

Figure 2. Basic steps of whey-ethanol production

THE ECONOMICS OF PRODUCING FUEL ETHANOL FROM WHEY PERMEATE

Estimated Cost of Producing Fuel Ethanol from Whey Permeate

With only two industrial-scale whey-to-ethanol plants in the United States, no publicly available production cost data exist. Costs quoted by several sources do not have enough details and probably represent the best "educated" estimates. However, a recent comprehensive cost estimate of producing fuel ethanol from whey is publicly available in a June 2005 New Zealand report (*Waste Solutions Ltd.*).

The New Zealand report related that the estimated "at-gate" cost (operating and capital service costs) of producing ethanol from whey permeate at maximum technical potential, with a level of uncertainty of +/- 20 percent, was N.Z. $0.6-0.7 per liter. Using a currency exchange rate of N.Z. $1 = U.S. $0.7, the estimated cost translated to U.S. $1.60-1.85 per gallon.

The estimated cost took into consideration economy-of-scale effects, transportation costs, and competing waste uses, and included the following scenario and assumptions:

- Fermentation at local plants.
- Distillation to 96-percent ethanol at local plants.
- Transportation of 96-percent ethanol to a centrally located dehydration plant.
- Capital service cost per year was 20 percent of capital cost, assuming a mixture of debt and equity financing and a nominal interest rate of 10 percent.
- To be economically feasible, the dehydration plant needed to have a minimum daily capacity of 60,000 liters of ethanol (about 15,850 gallons a day or 5 million gallons a year).
- For alcohol recovery (distillation and dehydration), biogas from effluent treatment was used as fuel. (Surplus steam from the dairy plant or cogeneration plant would also help.)
- Wet feedstock that had at least 15 percent (by weight) fermentable sugar content could pro duce ethanol that was 9-10 percent (by volume) of the fermentation beer. The resulting ethanol recovery cost could be less than N.Z. $0.2 per liter (U.S. $0.52 per gallon). (For a beer that contained 3-4 percent ethanol, the ethanol recovery cost was at least N.Z. $0.54 per liter (U.S. $1.42 per gallon).)

The cost of producing ethanol from whey permeate estimated by the New Zealand report, at U.S. $1.60- 1.85 per gallon, with a level of uncertainty of +/- 20 percent, is similar to the costs quoted by sources in the United States. Estimates from these U.S. sources yielded an operating cost of about $1 per gallon. In addition, there was a capital service cost of between $0.30 and $0.80 per gallon, which was calculated at an assumed rate of 20 percent of capital cost. The capital service cost would have been higher or lower if the assumed rate had been different. Capital cost (cost of the plant construction project) had quite a wide range, from $1.50 to $4 per annual gallon for a commercial operation, depending on the scale of the plant.

CONVERSION FACTORS USED IN THIS REPORT

- One pound of lactose consumed in fermentation yields 0.538 pound of ethanol (theoretical yield).
- 1 liter of ethanol equals 0.7924 kilogram of ethanol.
- 1 gallon equals 3.785 liters
- 1 kilogram equals 2.2046 pounds
- 1 gallon of ethanol weighs 2.9992 kilogram or 6.6121 pounds.

Table 5. Lactose Input and Estimated Feedstock Cost Per Gallon of Ethanol at Selected Yield Level

Ethanol yield (Percent of theoretical yield)	Lactose input (Pounds per gallon ethanol)	Feedstock cost per gallon ethano (For every $0.01 net lactose value per pound)
100%	12.29	$0.1229
95%	12.94	$0.1294
90%	13.66	$0.1366
85%	14.46	$0.1446

Cost of Whey Permeate as Feedstock

The amount of lactose needed to produce a gallon of ethanol depends on the level at which the lactose is consumed in fermentation and the efficiency of ethanol conversion. It was reported in the 1990s that commercial plants fermenting natural-strength whey could utilize greater than 95 percent of lactose with a conversion efficiency of 80-85 percent of the theoretical value (*Mawson*).

With technology advancement over the years and using higher concentration of lactose for fermentation, lactose consumption in fermentation is nearly complete at the two U.S. commercial plants. Presumably, the conversion efficiency is also higher than in the 1990s. The actual yield is proprietary information and is therefore not available.

For reference purposes, it was reported that conversion of lactose to ethanol at 85.5 percent to 91 percent efficiency (0.46 to 0.49 gram ethanol per gram lactose) could be obtained (*Dale*).

It would take 12.29 pounds of lactose to produce a gallon of ethanol, if the lactose is completely consumed in fermentation and ethanol conversion is 100 percent of the theoretical yield (Table 5).

The estimated operating cost of $1 per gallon of ethanol assumes that whey permeate used in ethanol fermentation is a free (no cost) feedstock. This assumption is valid when there is surplus whey to be disposed of by any least-cost means. However, when whey powder and lactose have found wider uses and have increased in value, the determination of the cost of whey permeate as feedstock for ethanol fermentation becomes more complicated.

To illustrate the calculation, use, for example, a May 2007 lactose price of $0.9370 per pound (Table 2). Subtracting from this price an estimated processor cost of $0.20 per pound for crystallizing and drying lactose, the net value of lactose was $0.7370. For every $0.01 net lactose value, the feedstock cost for fermentation would be $0.1229 per gallon of ethanol (Table 5). Given that the net value of lactose was $0.7370 per pound, the feedstock cost would amount to $9.06 per gallon of ethanol (($0.7370/$0.01)*$0.1229). If lactose consumption is less than complete in fermentation and ethanol conversion is less than 100 percent of the theoretical yield, then more than 12.29 pounds of lactose is required to produce a gallon of ethanol, and the feedstock cost would be even higher.

(No publicly available processor cost data for lactose production is available. The $0.20 per pound estimate is used, considering make-allowances of $0.1956 for dry whey in the Federal Milk Market Orders (*U.S. Department of Agriculture, Agricultural Marketing Service*) and $0.267 for skim whey powder in California's Stabilization and Marketing Plans for Market Milk (*California Department of Food and Agriculture)*).

Another illustration could use the lactose price prior to the run-up of whey products prices in 2006. The 2005 annual average price of lactose was $0.2012 per pound (Table 2). The net value of lactose after allowing for $0.20 processor cost would be $0.0012 per pound. The opportunity cost of lactose as feedstock for fermentation would have been $0.015 per gallon of ethanol (($0.0012/.$0.01)*$0.1229), or almost zero.

The calculation of the opportunity cost of lactose for ethanol fermentation is valid only if there are competing uses of the same lactose, such as manufacturing dry whey, lactose, or other whey products. If there is no such competition, then the whey permeate some- how has to be disposed of and the opportunity cost of lactose for ethanol fermentation is likely to be zero or even negative.

As shown in Table 3, an estimated 1,857 million pounds of lactose were used in various whey products in 2006, and an estimated 2,576 million pounds (58 percent of total available lactose) were unaccounted for. Most of the unaccounted-for volume was likely to have been disposed of as waste. Accordingly, the lactose in this surplus whey would not carry an opportunity cost had it been used as feedstock for ethanol fermentation.

Nationally, the surplus whey situation will persist for the foreseeable future. Although prices of whey products have more than doubled since June 2006 (Table 2), whey products production has remained rather constant (Table 6, *previous page*). For the first 6 months of 2007, cumulative production of dry whey and lactose, respectively, was 2.5 percent higher than the amount of the same period last year, but whey protein concentrate was 10.3 percent lower. The combined volume of the 3 products for the 6 months in 2007, at 1,156 million pounds, was only 0.9 million pounds more than the same period in 2006.

In the short run, whey products manufacture is limited by the available plant capacity and the production volume can increase only marginally. This is because current whey products plants were built to operate at or near capacity. New investment to expand the capacity will not happen unless the industry is convinced that the recent price hikes are a long-term trend.

Although nationally whey is still in a surplus situation, its availability is often localized (regionalized). The cost of using it as feedstock for ethanol fermentation is therefore site specific.

Table 6. Whey Products Production by Product and Month, United States, 2006-2007

Product and month	By Month			Cumulative		
	2006	2007	Change	2006	2007	Change
Dry whey, total[1]	*-- 1,000 pounds --*		*Percent*	*-- 1,000 pounds --*		*Percent*
Jan	88,391	96,145	8.8	88,391	96,145	8.8
Feb	89,695	90,232	0.6	178,086	186,377	4.7
Mar	100,953	98,713	-2.2	279,039	285,090	2.2
Apr	95,662	96,818	1.2	374,701	381,908	1.9
May	97,295	99,198	2.0	471,996	481,106	1.9
Jun	89,701	94,628	5.5	561,697	575,734	2.5
Jul	95,226			656,923		
Aug	91,547			748,470		
Sep	86,271			834,741		
Oct	87,434			922,175		
Nov	85,865			1,008,040		
Dec	92,306			1,100,346		
Lactose, human & animal						
Jan	64,539	65,064	0.8	64,539	65,064	0.8
Feb	56,311	59,671	6.0	120,850	124,735	3.2
Mar	62,165	64,975	4.5	183,015	189,710	3.7
Apr	63,402	63,083	-0.5	246,417	252,793	2.6
May	66,057	67,145	1.6	312,474	319,938	2.4
Jun	60,027	62,013	3.3	372,501	381,951	2.5
Jul	60,333			432,834		
Aug	64,225			497,059		
Sep	58,964			556,023		
Oct	62,296			618,319		
Nov	60,505			678,824		
Dec	59,832			738,656		
Whey protein concentrate, total						
Jan	37,162	33,055	-11.1	37,162	33,055	-11.1
Feb	34,436	29,432	-14.5	71,598	62,487	-12.7
Mar	37,766	35,038	-7.2	109,364	97,525	-10.8
Apr	37,525	33,188	-11.6	146,889	130,713	-11.0
May	37,373	33,436	-10.5	184,262	164,149	-10.9
Jun	36,190	33,672	-7.0	220,452	197,821	-10.3
Jul	35,704			256,156		
Aug	35,024			291,180		
Sep	34,581			325,761		
Oct	34,581			360,342		
Nov	33,116			393,458		
Dec	34,266			427,724		

Source: Dairy Products, September 2007, USDA National Agricultural Statistics Service.

[1] Excludes all modified dry whey products.

Economic Feasibility of Producing Fuel Ethanol from Whey Permeate

Whether it is economically feasible to produce ethanol from whey permeate is determined by the balance of the costs and the expected revenues. Key components on the cost side are:

- Feedstock cost: Cost of whey permeate (lactose) as input for ethanol fermentation.
- Operating cost: Labor, energy, supplies, repair and maintenance, depreciation, insurance, licensing fees, etc.
- Capital service cost: Annual cost is calculated at a rate of capital cost prescribed by the decisionmaker, based on the opportunity cost (interest cost) of capital and risk premium for undertaking the investment. It may be at 10 percent of capital cost, 15 percent, 20 percent, or some other rate.

On the revenue side, there are three main considerations:

- Ethanol price: About 90 to 95 percent of ethanol is sold under long-term contracts (6 to 12 months). Many of these contracts are fixed- price. The remaining amount is sold on the spot market, and the spot-market prices fluctuate according to market conditions (*Renewable Fuels Association*). *The Annual Energy Outlook 2007 with Projections to 2030* forecasts ethanol wholesale price (in 2005 dollars) to be $2.520 per gallon in 2007, $2.066 in 2008, $2.099 in 2009, $1.814 in 2010, and $1.742 in 2011. Thereafter, the long-term trend is for the price to be in the $1 .650 to $1 .720 range (*DOE Energy Information Administration*).
- Byproducts value: Effluent dried as feed, digested for methane gas, or used for other purposes, with positive or negative returns.
- Incentives: Currently (through December 31, 2010) there is a small ethanol producer Federal tax credit of $0.10 per gallon, up to 15 million gallons or $1.5 million per year, for a production facility with up to 60 million gallons of annual production capacity (26 U.S.C. § 40. The citation is intended for information only. Please consult tax professionals for specific tax treatments.) Various grants, loans, and other incentives are also offered by various Federal and State programs.

Depending on the magnitude of capital service cost and assuming whey permeate is a free feedstock that is converted to ethanol at an operating cost of $1 per gallon, the ethanol price must be higher than the total cost of producing it for the new investment in the ethanol plant to be economically feasible. Various production incentives may lower the price level required for economic feasibility.

Net returns from the ethanol enterprise should be measured against the profitability of making other whey or whey-derived products or of other uses of whey, to determine whether ethanol production is a more worthwhile undertaking. A further consideration should be which of the whey enterprises fits best with a cooperative's overall business strategy.

Economy of Scale

Because whey contains about 93.5 percent water and only 4.9 percent lactose, even a whey-ethanol plant of modest size requires a very large cheese operation to provide whey for feedstock. This fact destines the scale of whey-ethanol plants to be much smaller in size than present-day corn-ethanol plants and higher in capital cost per annual gallon.

A larger scale whey-ethanol plant would benefit from scale economy, where the capital cost increases proportionately less than the increase in plant size, resulting in lower capital cost per annual gallon. In a joint project studying the cost of producing ethanol from corn and lignocellulosic feedstocks, researchers at USDA and DOE used the following expression for scaling capital cost for equipment:

$$\text{New cost} = \text{Original cost x (new size/original size)}^{\text{exponent}}$$

The joint report cited USDA's value of the scaling exponent of 0.6 and DOE's average value of 0.63; both were within the range of 0.6 to 0.7 commonly cited in cost estimation literature (*McAloon,, et al*).

Another estimate based on 19 corn-ethanol plants built between 1996 and 2004 and ranging in size from 15 million to 50 million gallons per year yielded a plant scaling exponent value of 0.77 (*$(S\&T)^2$ Consultants , Inc., et al*).

Regardless of the value of the scaling exponent, as long as it is less than one, there would be scale economy for plants that lie within the relevant size range. Although this reference is to plants making ethanol from corn starch and lignocellulosic feedstocks, it is reasonable to expect that a similar economy of scale would apply to whey-ethanol plants.

A larger sized whey-ethanol plant would lower the per-gallon capital cost, but also would require a larger whey volume, either from a larger-sized cheese plant or from a group of cheese plants.

WHEY-ETHANOL PLANT SCENARIOS AND ROLES OF DAIRY COOPERATIVES

The New Zealand report suggested an ethanol plant with a minimum annual capacity of 5 million gallons. Assuming lactose is completely consumed in fermentation and ethanol is produced at the level of the theoretical yield, the plant would require 195,000 pounds of lactose a day as input.

A Cheese-Whey/Ethanol Complex

A cheese plant with a daily capacity of processing 4.5 million pounds of milk—about the size of some of the largest cheese plants in the United States—would generate four million pounds of whey to supply the required 195,000 pounds of lactose to the ethanol plant. To pump over whey permeate and save on transportation costs, the ethanol plant should be located close to the cheese plant.

The setups of the two U.S. whey-ethanol plants, at Corona and Melrose, fit this single cheesewhey/ethanol plant complex scenario.

Multi-Plant Coordination

For cheese plants of more modest sizes (typically located in the more traditional dairy regions), assembling whey permeate for ethanol production would require coordination among plants. For example, three plants, each with a daily capacity of processing 1.5-2 million pounds of milk into cheese, would yield the necessary amount of whey to supply lactose to the ethanol plant. At each cheese plant, whey would be ultrafiltered (deproteinated) and the permeate would be concentrated by reverse osmosis to about 20 percent solids (15 percent lactose) and then shipped to the ethanol plant for fermentation. To reduce whey permeate shipping costs, the ethanol plant should be located adjacent to the largest cheese plant among the three or where it is most logical and appropriate.

This coordination scheme, in fact, has been in practice by dairy cooperatives for whey handling, where several cheese plants condense their whey and then ship the condensed whey (deproteinated or otherwise) to a whey powder plant for drying. The same scheme could be used to coordinate whey handling among cheese plants to supply a whey-ethanol plant. Furthermore, such a coordination scheme could be expanded to allow future whey-ethanol plants to be of greater capacity in order to take advantage of the economy of scale.

Roles of Dairy Cooperatives

If a new whey-ethanol plant is proved to be economically feasible and were to be built, the enterprise might be organized according to these forms:

- An ethanol plant adjacent to a dairy cooperative's large cheese plant, similar to the setups of the two existing plants.
- An ethanol plant that ferments whey permeate assembled from several cheese plants of a dairy cooperative.
- An ethanol plant that ferments whey permeate assembled from several cheese plants. The coordination of whey handling may be among a cooperative's and other cooperatives' cheese plants, or among a cooperative's and other cooperative and non-cooperative entities' cheese plants. The coordination may be carried out by contract or organized as a joint venture. Small cheese plants looking for opportunities to add value to whey may be inclined to participate in such undertaking.

Some Specific Issues in Whey-Ethanol Production

Because of the composition of whey, there are some issues that are specific to whey-ethanol production (*Dale*):

- Whey and whey permeate concentrate are very susceptible to contamination and spoilage.
- Whey permeate concentrate is costly to transport (mostly water).
- The fermentation is susceptible to lactic contamination. The fermentation systems must be very carefully designed and operated, basically to food-grade cleanliness or, for some systems, even to aseptic standards.
- Because the calcium salts in the whey are "reverse soluble" —becoming insoluble at higher temperatures—scaling of the distillation column could be a problem or a cause for concern.
- The effluent is high in chloride. This limits the rate of application on fields if land-spreading is used. Just land-spreading of the effluent can be a major operating cost. There are two ideas for higher value products from the spent effluent: (1) a base for a sports drink if whey permeate concentrate is the substrate, and (2) a mineral salt block for animals if permeate mother liquor is the substrate.

Some Historical Lessons

In the United States, fuel ethanol production started in the late 1970s. During the 1980s, about 165 commercial plants (plants with more than 500,000 gallons annual capacity) were constructed, with grain as the primary feedstock. By the end of 1990, fewer than 40 plants remained in operation (*Murtagh, et al*), although annual ethanol production grew to 900 million gallons (*Renewable Fuel Association*).

The reasons for the high attrition rate of plants during that decade were reviewed in a 1991 paper (*Murtagh, et al*). Experiences of that period may be relevant to future commercial whey-ethanol development and are summarized in this section. They may provide some useful lessons that illustrate the kind of mistakes to avoid.

The most significant causes of project failures during the 1980s were improper technology selection and improper engineering design. Every aspect of the plant operations was susceptible to such failures. From feedstock pretreatment to yeast propagation, fermentation, distillation, DDGS (distillers dried grains with solubles) drying and storage, and piping, the culpable factors were inadequate design, equipment, and/or process. Without being supported by adequate research, novel steps taken to save cost, increase yield, or otherwise cut corners, tended to invite disastrous results.

Other factors that contributed to failures were shifting public policy; fraudulent investment schemes; plants that were constructed with high cost, without feasibility studies, or without adequate financing; and lack of technical and managerial expertise in the biochemical process.

Conclusions

There is a potential for supplementing the Nation's fuel ethanol supply by an estimated 203 million gallons a year (2006 data) if all lactose in surplus whey and whey permeate—whey that is not used in value-added whey-derived products—is fermented for the purpose.

Dairy cooperatives could have a share of 65 million gallons of this potential. However, there are only two commercial whey-ethanol plants with an annual production of 8 million gallons. Both plants are currently owned and operated by dairy cooperatives.

The fact that the two plants have been in operation for more than 20 years is an indication that (1) fuel ethanol production from whey is technically feasible, (2) whey-to-fuel ethanol production technologies and processes are mature and capable of being adopted for commercial operations, and (3) producing fuel ethanol from whey is economically feasible.

Because there are no publicly available, actual production-cost data, no attempt was made to estimate the profitability of the whey-to-ethanol enterprise. The cost of producing ethanol from whey permeate estimated by the 2005 New Zealand report was U.S. $1.60-1.85 per gallon (at a currency exchange rate of N.Z. $1 = U.S. $0.7), with a level of uncertainty of +/- 20 percent. Estimates ascertained from U.S. sources in the course of this study yielded a per-gallon operating cost of about $1 and a capital service cost that may be calculated on a capital cost of $1.50 to $4 per annual gallon. These cost estimates have a wide range of uncer tainties and are also sensitive to the scale of the plant.

Then there is also the uncertainty regarding the cost of using whey permeate as feedstock. Prior to the price run-up in 2006, the dairy industry's main task concerning whey had been to seek more methods for whey to be useful and valuable. Under those circumstances, whey and whey permeate used in fermentation may be regarded as a feedstock of no or even negative cost. This free feedstock premise remains true if there are no readily accessible, profitable alternatives for the whey.

In assessing the feasibility of a new whey-ethanol plant, the cost of whey permeate as feedstock needs to be carefully evaluated in this era of whey products' price uncertainties. Every 1 cent of net lactose value from alternative uses would increase the fermentation feedstock cost by at least 12.29 cents per gallon of ethanol. Other important factors to consider besides feedstock cost are (1) an appropriate plant scale that would minimize capital cost and the cost of assembling feedstock, (2) an appropriate technology and process specifically for whey-ethanol production that would minimize operating cost, (3) best alternatives for using and/or disposing of the effluent, (4) ethanol price, and (5) various government production incentives.

The *OECD-FAO Agricultural Outlook 2007-2016* provides the only available long-term dry whey price projection. It forecasts the wholesale price of edible dry whey (F.O.B., Wisconsin plant) to peak in 2011, but the decline afterwards will still see the 2016 price to be 20 percent higher than in 2006 (*OECD-FAO*). On the other hand, the *Annual Energy Outlook 2007 with Projections to 2030* forecasts the ethanol wholesale price to peak in 2007 ($2.520 per gallon) and fluctuate in the $1.650 to $1.720 range after 2011 (*DOE Energy Information Administration*). However, care should be used if the two projected price series are to be employed for evaluating the feasibility of a new whey- ethanol plant versus a new dry whey plant, because the forecast of the dry whey price is in nominal dollar and the ethanol price is in constant (2005) dollar. Further complicating the picture is that the projection of the dry whey price preceded, and therefore did not incorporate, the unexpected price surges in 2007.

References

Alexander, Craig & Nelson, Mark. (2002). *An Economic Analysis of the US Market for Lactose*, Cornell Program on Dairy Markets and Policy, presented at Cornell Conference on Dairy Markets and Product Research, March.

Anchor Ethanol. http://www.nzmp.com/cda/frontpage/0,,c334829 g500040,00.html.

Audic, J. L., Chaufer, B. & Daufin, G. (2003). "Non-food Applications of Milk Components and Dairy Co-products: A Review," *Le Lait (Dairy Science and Technology) 83* pp. 471-438, EDP Sciences.

Belem, M. A. F. & Lee, B. H. (1998). "Production of Bioingredients from Kluyveromyces Marxianus Grown on Whey: An Alternative," *Critical Reviews in Food Science and Nutrition*, Volume *38*, Number 7, October, pp. 565-598.

Bio-Process Innovation, Inc. "Whey Lactose Ethanol- BPI Technologies for Whey Ethanol," http://bioprocess.com/wheyEthanol.html.

California Department of Food and Agriculture, Dairy Marketing Branch. (2006). Hearing Decision Notice, July 24.

Chandan, Ramesh. (1997). *Dairy-Based Ingredients*, Eagan Press, St, Paul, MN.

Cotanch, K. W., Darrah, J. W. Miller, T. K. & Hoover, W. H. (2004). *"The Effect of Feeding Lactose in the Form of Whey Permeate on the Productivity of Lactating Dairy Cattle: Final Report,"* W. H. Miner Agricultural Research Institute, Chazy, NY, and Rumen Fermentation Profiling Laboratory, West Virginia University, February 23.

DAIRY FACTS, Issue 96, March (2007), *"High Tech Cheese Plant Comes to Clovis, N.M.,"* Department of Food Science & Technology, Virginia Polytechnic Institute and State University, http://www.fst.vt.edu/extension/drg/dfax/Mar07.html.

Dairy Farmers of America (DFA). (2007). *News Release*, "Dairy Farmers of America Announces Changes to American Cheese Division," August 8.

Dale, M. C. (2007). *Discussions on Bio-Process Innovation*, Inc., processes, e-mail correspondence, April 15.

Dale, M. C. & Moelhman, M. (1997). *A low-energy continuous reactor-separator for ethanol from starch, whey permeate, permeate mother liquor, molasses or cellulosics.* Project final report, April 1, 1994-February 28. U.S. Department of Energy, DOE/CE/15594-T11, April 14, 1997. http://www.osti.gov/energycitations/servlets/purl/4691 82-RsHsi0/webviewable/469182.pdf.

Desmond, John. (2007). *A brief description of the Carbery process,* e-mail correspondence, May 15.

Farmnews for NZ Farmers. "First Milk-Made Fuel Poured Today, " http://www.farmnews.co.nz/news/2007/aug/775.shtml.

Hamilton, Ron (AnchorProducts, Tirau). (2006). *"The Manufacture of Ethanol from Whey,"* New Zealand Institute of Chemistry (modified on May 22). http://www.nzic .org.nz/ChemProcesses/dairy

Irish Examiner.com. "Maxol launches biofuel mix for standard vehicles," http://www.examiner.ie/story/?jp=EYKFKFKFSN&cat= Business&rss=rss2.

Kosikowski, Frank V. & Mistry, Vikram V. (2004). *Cheese and Fermented Milk Foods*, Volume I, Origins and Principles, third edition, F. V. Kosikowski, L.L.C., Westport, Connecticut, 1997, Chapter 26. Ling, K. Charles. *Marketing Operations of Dairy*

Cooperatives, 2002, USDA Rural Business-Cooperative Service, RBS Research Report 201, February.

Mackle, Tim. "BioEthanol as a Transport Fuel," Anchor Ethanol, Fonterra Cooperative Group Ltd. http://eeca.govt.nz/eeca-library/renewable-energy/

Mawson, A. J. (1994). "Bioconversions for Whey Utilization and Waste Abatement," *Bioresource Technology 47:* 195-203.

McAloon, Andrew, Frank Taylor, & Winnie Yee (2000). (U.S. Department of Agriculture), and Kelly Ibsen, and Robert Wooley (Department of Energy). *Determining the Cost of Producing Ethanol from Corn Starch and Lignocellulosic Feedstocks*, DOE National Renewable Energy Laboratory, Technical Report NREL/TP-580- 28893, October.

Murtagh, John E. "*Commercial Production of Ethanol from Cheese Whey - The Carbery Process,*" Energy from Biomass and Wastes IX, Symposium, 1/28- 2/1/1985, Buena Vista, FL, Institute of Gas Technology, pp. 1029-1039.

Organization for Economic Co-operation and Development (OECD) and Food and Agriculture Organization of the United Nations (FAO). *OECD- FAO Agricultural Outlook 2007-2016.*

Renewable Fuels Association. "Industry Statistics," http://www.ethanolrfa.org/ industry

Roehr, M. (edited). (2001). *The Biotechnology of Ethanol: Classical and Future Applications*, Wiley-VCH,, Part II, Chapter 5, pp. 139-148.

Sandbach, D. M. L. (1981). "Production of Potable Grade Alcohol from Whey," *Cultured Dairy Products Journal, American Cultured Dairy Products Institute, Volume 16 (4),* November, pp. 17-22.

$(S\&T)^2$ Consultants, Inc., and Meyers Norris Penny LLP. (2004). *Economic, Financial, Social Analysis and Public Policies for Fuel Ethanol: Phase I*, Prepared for Natural Resources Canada, Office of Energy Efficiency, November 22.

U.S. Department of Agriculture, Agricultural Marketing Service. (2006). "Federal Milk Marketing Orders, Class III/IV Price Make-Allowances, Interim Final Rule," *Federal Register*, December 29.

U.S. Department of Agriculture, Economic Research Service. (2006). *Livestock, Dairy, & Poultry Outlook*, various issues since October 19.

U.S. Department of Energy, Energy Information Administration. (2007). *Annual Energy Outlook 2007 with Projections to 2030*, Report #: DOE/EIA-0383, February 2007, Table 12 (Petroleum Product Prices).

U.S. Department of Energy, Office of Industrial Technologies, Energy Efficiency & Renewable Energy. (2001). *Continuous Cascade Fermentation System for Chemical Precursors: Low-Energy Continuous System Coverts Waste Biomass to Ethanol*, Order # 1-AG-594, September.

U.S. Patent & Trademark Office. (1939). United States Patent 2,183,141, Kauffmann, W., and van der Lee, P. J. "*Method of Fermenting Whey to Produce Alcohol*," December 12.

U.S. Patent & Trademark Office. (1987). United States Patent 4,665,027, Dale, et al. "*Immobilized Cell Reactor-Separator with Simultaneous Product Separation and Methods for Design and Use Thereof*," May 12.

The Maxol Group. "Moxol Bioethanol E85Join the Alternative Fuel Revolution," http://www.maxol.ie/E85/index.html.

Waste Solutions Ltd. (2005). *Estimate of the Energy Potential for Ethanol from Putrescible Waste in New Zealand*, Technical Report prepared for the Energy Efficiency and

Conservation Authority, June. http://www.eeca.govt.nz/eeca-library/renewable-energy/biofuels/report/energy-potential-for-fuel-ethanol- from-putrescible-waste-in-nz-report-05.pdf.
Webb, B. H. and E. O. Whittier. *Byproducts of Milk*, second edition, Avi, Westport, CT, 1970, table 14.1, p. 408.
Wendoff, W. L. (1993). "Revised Guidelines for Landspreading Whey and Whey Permeate," Department of Food Science, University of Wisconsin- Madison, UW DAIRY ALERT! June 1.
Wisconsin Center for Dairy Research. *Approximate Distribution of Milk Components between Cheese and Whey, and in Whey Products*. http://www.cdr.wisc.edu/applications/ whey/
Yang, S. T. & Silva, E. M. (1995). "Novel Products and New Technologies for Use of a Familiar Carbohydrate, Milk Lactose," *Journal of Dairy Science 78*: 2541-2562, table 8.

APPENDIX I. THE CARBERY PROCESS

The Carbery process in its present-day operation is provided by the Carbery Group (formerly Carbery Milk Products Ltd.), Ballineen, County Cork, Ireland (*Desmond*, used with permission).

"In the alcohol plan the raw material, whey permeate, is converted into finished product, potable alcohol. The first step in this process is the conversion of the carbohydrate in the permeate (lactose) into ethyl alcohol. This is achieved by fermentation with a specific yeast strain.

The fermentation is carried out in eleven cylin dro-conical fermenter vessels. Compressed air is used to agitate the contents of the vessel and it also provides aeration of the contents to encourage continued yeast growth.

Whey permeate and yeast are added together into a fermenter and the fermentation is allowed to proceed under the optimum conditions of temperature, pressure and agitation, until all of the lactose in the permeate has been exhausted. The lactose is converted mainly into ethyl alcohol, but other compounds known as congeners are also produced by the fermentation. Depending on the initial lactose concentration and yeast activity, the fermentation will take between 12 and 20 hours to complete.

After fermentation, the contents of the fermenter are referred to as 'wash' or 'beer.' The alcohol content of the wash will depend upon the initial lactose concentration in the permeate and the fermentation efficiency. The wash is pumped to the distillation plant.

The next operational step performed on the wash is distillation. The purpose of the distillation step is to concentrate the alcohol portion of the wash, and to remove the congeners formed during fermentation. A continuous-distillation process employing column stills is used. It consists of three sections:

(i) Beerstill.
(ii) Extractive-distillation unit.
(iii) Rectifier.

In the beerstill, the wash is concentrated to 96 percent alcohol. This is then fed to the extractive-distillation column where water is added, changing the boiling point of the mixture, so that highboiling-point 'higher alcohols' may be removed. Finally, in the rectifier, the alcohol strength which has been reduced in the extractive-distillation column is increased again to 96 percent. Other congeners such as 'heads', 'esters' and 'fusel oil' are also removed in this final rectifier.

Most distillation units consist of a cylindrical, vertical column. Perforated plates (sieve trays) are fixed horizontally at intervals of several inches throughout the height of the column. Liquid is usually introduced to a plate approximately half-way up the column. Steam is introduced at the base. The steam and vaporized liquid tend to rise up the column through the plate perforations, while the liquid tends to fall to the bottom via a series of down pipes.

Alcohol with a boiling point of 78°C is more volatile than the water portion. The alcohol will tend to rise up the column in the vapor whereas the water will tend to go down the column with the liquid. The alcohol is concentrated to 96 percent in a concentrating column. The product is removed to storage/further rectification, and spent wash, which contains very little alcohol, is removed from the bottom of the column.

Vapors rising above the top plate of the col umn are condensed in one or more condensers, and a reflux line returns the condensate to the uppermost plates, above the point where the product is drawn off, thus maintaining a liquid level on the draw tray."

Ardent readers will find the process has evolved over the years when compared with the original setup (*Sandbach)*. It also should be noted that potable alcohol and fuel ethanol have different quality requirements, and the processes of producing them may differ somewhat, although the basic principles are the same.

In addition, for potable alcohol, ethanol concentrations post-fermentation typically range from 2.5 percent to 3.5 percent (*Desmond*). Fuel ethanol production requires ethanol content in the beer to be at least double that level for energy-efficient distillation and dehydration.

APPENDIX II. THE PROCESSES OF BIO-PROCESS INNOVATION, INC.

The processes offered by Bio-Process Innovation, Inc. (BPI), are the culmination of many years of research (e.g., *Dale, et al*). Depending on the feedstock and efficiency desired by a plant, the company offers four kinds of fermentation systems (*Bio-Process Innovation, Inc.*):

1. **Immobilized Cell Reactor/Separator (ICRS)**
   This patented immobilized cell reactor/separator separates ethanol as it is being produced and allows the quick and continuous conversion of clear whey permeate concentrate to ethanol. The experiment for the patent application (*U.S. Patent no. 4,665,027*) shows that it has an initial inlet lactose concentration of about double the natural lactose content in whey permeate and a sugar utilization rate of 98 percent. The outlet (effluent) BOD is about 2.5 percent of its original value.

2. **Continuous Stirred Reactor/Separator**
   This technology allows high rates of fermentation coupled with ethanol recovery from the fermentation vessel. Yeast can be immobilized or recycled to keep fermentation rates high.

3. **Continuous Cascade Reactor**
   Three or four stage continuous cascade fermentation system coupled with the company's proprietary salt and ethanol tolerant strains of *Kluyveromyces marxianus* (a lactose fermenting yeast family) allows 7 to 10 percent ethanol to be made from permeate mother liquor, whey permeate concentrate, or lactose. (*U.S. Department of Energy, Office of Industrial Technologies, Energy Efficiency and Renewable Energy* highlights the Continuous Cascade Reactor as a low-energy continuous system for converting waste biomass to ethanol.

4. **Batch Fermentation of Permeate Mother Liquor**

Actual performance for the fermentation systems two through four might be different from Immobilized Cell Reactor/Separator (system 1). But each of the four systems can attain near complete conversion of lactose to ethanol, at 0.46 to 0.49 gram ethanol per gram lactose. Outlet ethanol will be high if the process does not include simultaneous separation; low, if the technology is incorporated (*Dale*).

These associated technologies work with the above systems: (1) salt and ethanol tolerant strains of *Kluyveromyces marxianus*, (2) yeast production from whey permeate, and (3) low energy/non-fouling distillation of beers produced from whey permeate or permeate mother liquor.

USDA Rural Development
Rural Business and Cooperative Programs
Stop 3250
Washington, D.C. 20250-3250

USDA Rural Development provides research, management, and educational assistance to cooperatives to strengthen the economic position of farmers and other rural residents. It works directly with cooperative leaders and Federal and State agencies to improve organization, leadership, and operation of cooperatives and to give guidance to further development.

The cooperative segment of USDA Rural Development (1) helps farmers and other rural residents develop cooperatives to obtain supplies and services at lower cost and to get better prices for products they sell; (2) advises rural residents on developing existing resources through cooperative action to enhance rural living; (3) helps cooperatives improve services and operating efficiency; (4) informs members, directors, employees, and the public on how cooperatives work and benefit their members and their communities; and (5) encourages international cooperative programs. Rural Development also publishes research and educational materials and issues *Rural Cooperatives* magazine.

sex, marital status, familial status, parental status, religion, sexual orientation, genetic information, political beliefs, reprisal, or because all or a part of an individual's income is derived from any public assistance program. (Not all prohibited bases apply to all programs.) Persons with disabilities who require alternative means for communication of program information (Braille, large print, audiotape, etc.) should contact USDA's TARGET Center at (202) 720-2600 (voice and TDD). To file a complaint of discrimination write to USDA, Director, Office of Civil Rights, 1400 Independence Avenue, S.W., Washington, D.C. 20250-9410 or call (800) 795-3272 (voice) or (202) 720-6382 (TDD). USDA is an equal opportunity provider and employer.

CHAPTER SOURCES

The following chapters have been previously published:

Chapter 1 - This is an edited, reformatted and augmented version of a U.S. Department of Agriculture publication, Agricultural Economic Report Number 842, dated February 2007.
Chapter 2 – This is an edited, reformatted and augmented version of National Renewable Energy Laboratory Publication NREL/BR-510-40742, dated March 2007.
Chapter 3 - This is an edited, reformatted and augmented version of National Renewable Energy Laboratory Technical Report NREL/TP-510-41168, dated April 2007.
Chapter 4 - This is an edited, reformatted and augmented version of a U. S. Department of Agriculture publication, Research Report 214, dated February 2008.

INDEX

A

B

C

D

E

F

I

J

K

L

R

S

T

U